Asian-Pacific Regional Security

The Washington Institute for Values in Public Policy
The Washington Institute sponsors research that helps provide the
information and fresh insights necessary for formulating policy in
a democratic society. Founded in 1982, the Institute is an independent, non-profit educational and research organization which
examines current and upcoming issues with particular attention to
ethical implications.

ADDITIONAL TITLES

*Soviet Nomenklatura: A Comprehensive Roster of Soviet Civilian and
Military Officials (Third edition, revised and updated)*
Compiled by Albert L. Weeks (1990)

*The Nixon Kissinger Years:
The Reshaping of American Foreign Policy*
By Richard L. Thornton (1989)

Vietnam: Strategy for a Stalemate
By F. Charles Parker (1989)

*The Politics of Latin American Liberation Theology:
Challenges to U.S. Public Policy*
Edited by Richard L. Rubenstein and John K. Roth (1988)

*The Dissolving Alliance: The United States
and the Future of Europe*
Edited by Richard L. Rubenstein (1987)

Arms Control: The American Dilemma
Edited by William R. Kintner (1987)

*The East Wind Subsides: Chinese Foreign Policy and the Origins of the
Cultural Revolution*
By Andrew Hall Wedeman (1987)

Rebuilding a Nation: Philippine Challenges and American Policy
Edited by Carl H. Landé (1987)

Asian-Pacific Regional Security

Edited by June Teufel Dreyer

Published in the United States by The Washington Institute Press
Suite 300, 1015 18th Street, NW, Washington, D.C. 20036

Cover design and maps by Paul Woodward
Page design and typesetting by Edington-Rand, Inc.

Library of Congress Cataloging-in-Publication Data
Asian Pacific regional security / edited by June Teufel Dreyer
    p.   cm.
    Includes bibliographical references and index.
    ISBN 0-88702-053-4 (softcover): $14.95.
    1. Pacific Area—National security. 2. East Asia—National
security. 3. United States—Military relations—Pacific Area.
4. Pacific Area—Military relations—United States. 5. East Asia—
Military relations—United States. 6. United States—Military
relations—East Asia. I. Dreyer, June Teufel, 1939–
UA830.A8348 1990
355′.03301823—dc20   90-12832
                                                      CIP

Table of Contents

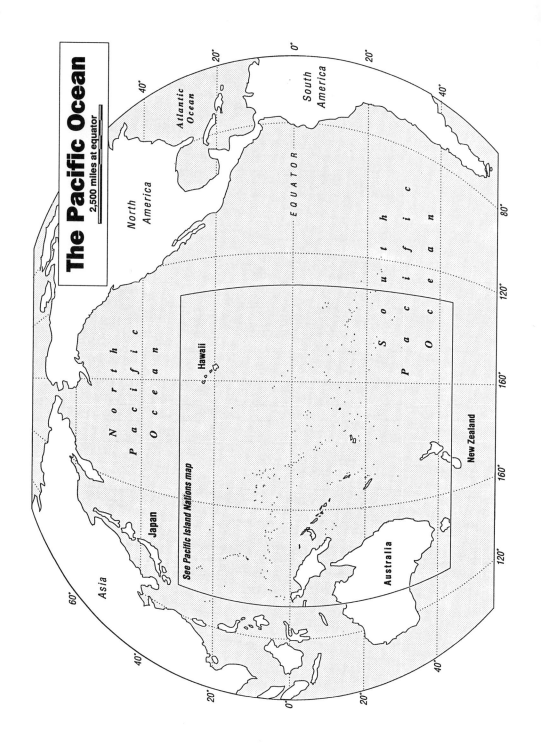

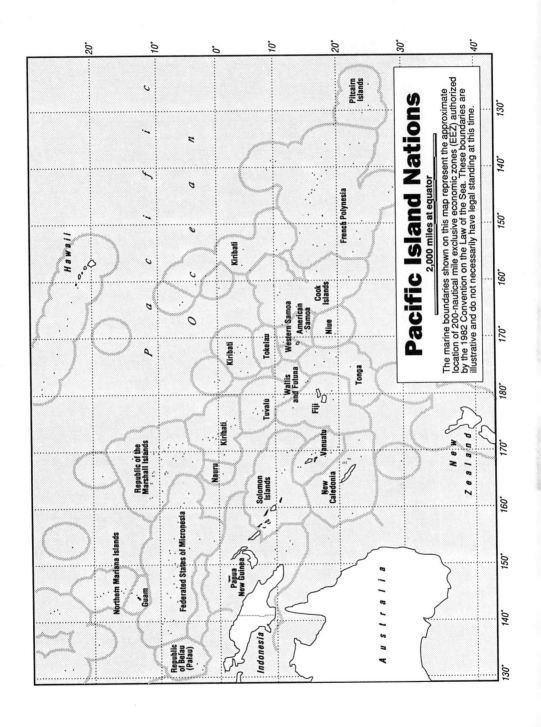

Pacific Island Nations

2,000 miles at equator

The marine boundaries shown on this map represent the approximate location of 200-nautical mile exclusive economic zones (EEZ) authorized by the 1982 Convention on the Law of the Sea. These boundaries are illustrative and do not necessarily have legal standing at this time.

Pacific Ocean

Hawaii

Northern Mariana Islands

Guam

Republic of Belau (Palau)

Federated States of Micronesia

Republic of the Marshall Islands

Kiribati

Kiribati

Kiribati

Nauru

Papua New Guinea

Indonesia

Solomon Islands

Tuvalu

Vanuatu

New Caledonia

Wallis and Futuna

Fiji

Tokelau

Western Samoa

American Samoa

Cook Islands

Niue

Tonga

French Polynesia

Pitcairn Islands

Australia

New Zealand

20° 10° 0° 10° 20° 30° 40°

130° 140° 150° 160° 170° 180° 170° 160° 150° 140° 130°

Foreword

Senator Richard G. Lugar

The openness of United States markets and the forward presence of American military forces have contributed significantly to the peace and prosperity enjoyed by the countries of the Pacific Rim for many years. No vacuum of military authority has been countenanced, and thus no major threat to the territorial integrity of nation-states in the region has emerged. Over sealanes kept open without visible threats to shipping, Asian countries have increased exports to a relatively open United States market in a dynamic fashion that has contributed to their sustained economic growth. Any review of the security equation in Asia and the Pacific must begin with and underline the importance of American openness and forward presence to the fulfillment of many of the aspirations of the Pacific Rim nations over the last two decades.

The decade of the 1990s, however, will commence with both American openness and its forward military presence in the Asian-Pacific region increasingly subject to vociferous debate at home and abroad. Even as the Uruguay round of GATT lurches toward a climax in the 1990s—hopefully with progress recorded in the area of general reductions in tariff barriers or at least an explicit recognition of non-tariff barriers as a prerequisite to systematic reduction—the United States trade deficit with the rest of the world will likely remain well above $100 billion. Our cumulative external debt may well be in excess of a half-trillion dollars. These tendencies, along with the more specific but highly publicized annual trade deficit with Japan, in excess of $50 billion, and the smaller but still notable deficits with Taiwan and South Korea, have stimulated a significant alteration in the United States domestic political climate.

Occasional public opinion polls declare that Japan is perceived by a majority of Americans as more threatening than the Soviet Union. Various authors attempt to exploit a demand for comprehensive knowledge of the Japanese with tomes that describe a predatory mercantilist power with a culture and ideas about democracy very different from those entertained by Americans. The debate in the United States Senate on the FSX co-production agreement with Japan lifted political passions and economic techno-nationalism to new heights and cut to the heart of the basic Japanese–United States bilateral relationship.

In other Senate debates during the 101st Congress, the United States troop commitment to South Korea has been questioned. The U.S. Congress has warmly applauded the administration of President Corazon Aquino of the Philippines while worrying aloud about some of the demands heard in the Philippine Congress regarding negotiations on the future of Clark Air Field and Subic Bay Naval Base.

Although the Soviet increase in naval and other military forces in the Pacific over the past decade has shown no signs of abatement, President Mikhail Gorbachev's suggestion in 1988 of mutual reductions in bases and forces by the United States and the Soviet Union has served to further complicate matters. Moreover, current negotiations in Vienna between members of NATO and the Warsaw Pact over reducing military personnel in the heart of Europe and the destruction of thousands of tanks, aircraft, and other "conventional" weapons have encouraged some along the Pacific Rim to call for an Asian negotiating equivalent. When coupled with an appreciation of European efforts toward greater unity and harmony in 1992, it is somewhat understandable that some political leaders in various Asian and Pacific countries are increasingly questioning the postwar economic and security order in the region.

Before such tendencies become too pronounced or are translated into policy prematurely, some of us who recognize the gains in the areas of peace and prosperity achieved by the countries of the region and, of equal importance, the foundations behind those gains, should speak out strongly and unambiguously.

And our message, delivered in unison as much as possible, must be that security in Asia and the Pacific will continue to depend for the foreseeable future on the forward presence of the American military and the openness of the American markets to exports from the region. While the validity of this observation may appear self-evident to some, it is, unfortunately, too often taken for granted.

As the Republican floor leader during the two Senate debates on the FSX aircraft co-production agreement with Japan, I mentioned a conversation that I enjoyed with Prime Minister Lee Kuan Yew of Singapore in Jaunary 1989. Lee stated that under no circumstances should the United States military presence leave Asia. His comments coincided with those of Japanese parliamentarians who reiterated the Japanese constitutional limits on the rebuilding of Japanese military power.

Lee argued that without United States military protection, Japan would be compelled to protect itself. In due course, Japan would find a manifest destiny to fill the vacuum left by American withdrawal and insist on protecting the ASEAN nations. From time to time, the nations involved may comfort themselves that the Soviet Union is withdrawing or has lost interest or that the Chinese are preoccupied with internal matters, but such diminished interest or preoccupation may not continue following United States withdrawal from a geographical region that includes China and the Soviet Union.

So strongly did Lee wish to make his point that he offered to discuss potential support for the United States Navy in Singapore and potential support for the United States Air Force in other regional circumstances. In talks with other Asian leaders in 1989, I noted a constant refrain of doubt about the physical staying power of the United States military in Asia and open speculation about the consequences of a power vacuum and about potential successors to fill that vacuum.

United States domestic political consideration of our role in Asian regional security is made increasingly difficult by a seemingly endless series of arguments on trade issues. Governments usually try to separate security issues from trade issues and to conduct negotiations in separate compartments. But average American citizens make no such separation, nor do most of their elected representatives.

I am not the only American who has explained tactfully and often to leaders in Japan, South Korea, Taiwan, Thailand, and Singapore that the trade surpluses that each of those countries continues to maintain with the United States are dangerously unsustainable, both economically and politically. They are dangerous because they may contribute toward a severe recession in the United States that will likely cut United States imports and trade deficits dramatically and produce severe economic crises in the economies of our trading partners.

These deficits are dangerous because they will increase pressure for American military withdrawal from the Pacific to save money.

They are dangerous because they contribute to endless disputes in Asian countries about the United States military presence and alleged United States bullying as we assert our trade interests. Many Asian politicians, business persons, students, and self-proclaimed nationalists conclude that Asia would be happier with fewer and less forward Americans.

Many South Korean students seriously contend that the United States military presence prevents peaceful reunification of North and South Korea and props up the current democratically elected government. Many Filipino politicians and intellectuals contend that "manhood" for the Philippines is impossible until the United States military withdraws from Clark and Subic Bay. Japanese businessmen speak and write about the whining excuses of an uncompetitive American business sector populated by management and workers who lack self-discipline, ambition, and adequate education.

An increasing number of Americans are fully prepared to respond in kind with enthusiastic calls for the removal of all military presence in the Asian sphere and a thoroughly emotional call for large-scale shutouts of Asian exports. To the extent that American and Asian leaders are swept along by these sentiments of withdrawal, protectionism, and mutual recrimination, security in Asia and the Pacific will decline dramatically. How ironic and tragic this would be! By all historic measurements we have erected a remarkable edifice of security and prosperity that should be envied, celebrated, and used as the foundation for improvements in the future.

For the moment, the co-development phase of the FSX aircraft agreement with Japan moves forward. It is critical that it should result in production of an aircraft that will meet the Japanese security needs in patrolling strategic straits and naval passages.

For the foreseeable future, the Soviet Pacific fleet is a threat to Japan. One need not disparage Soviet efforts to mend relations with Japan at a time of tensions in the United States–Japanese bilateral relationship in order to justify the continuing need for United States–Japanese cooperation in the naval field. A Soviet peace treaty with Japan that resolved the issue of the disputed Northern Islands would likely have little impact on Japan's security equation or its position in the Pacific. While *glasnost* and *perestroika* have had some far-reaching foreign policy implications in Eastern and Western Europe, the impacts of these reforms in the Pacific are not as clear. The Japanese are under no security illusions.

Less well understood is the delicate relationship among American military protection, the rebirth in Japan of desires to build aircraft and other military instruments independently, and the conduct of an aggressive Japanese export policy that threatens to destabilize not only Japan's relationship with its best market but also with many other trading partners vital to its economic security. For the moment, the FSX debate has served both to define some limits to the Japanese go-it-alone movement and to simultaneously rekindle discussion of the advisability of proceeding with such a policy in the future. A strategy for containing the trade imbalance is apparent neither in the United States nor in Japan, but it must be a primary goal of top leaders of both countries. In my judgment, the security relationship is in some jeopardy without resolution of the trade imbalance.

The result of the United States–Philippines base negotiations is difficult to predict, but I am hopeful that the outcome will include a continued United States military presence, a strong new surge of cooperation between the United States and the Philippines in behalf of mutual security objectives, and strong new ties with the ASEAN nations for mutual security in the area. The agreement should be for the long term and not a formula for incremental withdrawal. Already, creative United States and Filipino private-sector leaders have sketched out ways in which military and economic security needs for the Philippines might be met simultaneously in a long-term base agreement. The external debt of the Philippines and the need for substantial new capital investment could be addressed. Filipino democracy and nationalism must be carefully considered and respected.

The United States played an important role in the birth of Filipino democracy. The potential for a long and productive new friendship based upon mutual respect, admiration, and appreciation is possible, although clearly not inevitable. If Clark and Subic Bay are closed to the United States, we may have to work out substitute arrangements with other nations. The change will mark a net loss for the Philippines and for security in Asia and the Pacific generally. To the extent that such a change is expensive, it may jeopardize a portion of the United States security contribution to Asia.

President Bush has counseled a "wait and see" posture on developments in the People's Republic of China. For the moment, it is clear that the Chinese leadership has chosen survival and continuation first and foremost, above any other security or economic objectives. But China is weaker after this new wave of repression. Should stability in China be further undermined by the rise of competing regional power centers throughout the country, she might become somewhat more threatening to her immediate neighbors. For the moment, China's relative weakness probably enhances the potential for greater Soviet adventurism directed toward Japan or South Korea with North Korean assistance.

The Chinese situation highlights the end of British control in Hong Kong and the potential for an even more horrendous refugee exodus on top of the extraordinary refugee movements already manifest in Southeast Asia.

The United States and Canada have accepted tens of thousands of refugees from Asia. This policy has triggered another domestic debate in the United States on immigration policy, with increasing attention paid to the percentages and numbers of immigrants from the Pacific Rim.

The United States has gained immeasurably from the flow of human capital into our country. We should provide for expanded gateways of opportunity. In much the same way that economists now point to the increased potential for vigor in the West German economy occasioned by the entry of East German refugees, the surge of Asian immigrants to the United States may prove to be the bridge that our country needs to increase productivity with a more skillful workforce at a time of adverse demographic changes. The confidence of our friends in Asia and the Pacific will be enhanced by our vigor in meeting the needs of refugees who are fleeing despotism and seeking democratic governance and a free market system. We have gained immensely by educating tens of thousands of Chinese students in recent years and offering a home to 40,000 now engaged in studies in the United States.

Political observers often note the fragility of growing democratic institutions in the Philippines, South Korea, Taiwan, and Singapore and question our seeming preoccupation with support for democracy in an area in which *Realpolitik*, with or without democracy, might appear more in order.

But security in Asia and the Pacific is built upon a strong movement by all of the people involved toward the exercise and enjoyment of human rights and the building of democratic institutions that encourage and protect religious, political, economic, and academic freedom. The openness of these systems guards against unpleasant and deadly surprises. Perhaps equally important, the American people simply do not wish to support regimes in which authoritarian leaders suppress people. For many years, some American leaders argued that the crucial factor in determining the level of United States support was the fervor of anticommunism expressed by foreign governments who, in an imperfect and dangerous world, might not have any democratic pretensions. The American people have now decided that anticommunism alone is not a sufficient reason for American support. Domestic debates of the future in the United States are unlikely to result in substantive support for regimes that are not engaged in substantial democracy building.

I share with the authors of this book the fundamental view that Asia and the Pacific are tremendously important to the United States. Our failure to make strong relationships work would undermine the security and the prosperity of our country. I also share the common assessment that we have enjoyed a great period of growth together in the Pacific and that rarely have we witnessed such security and prospects for a peaceful and prosperous future.

Perhaps the day-to-day arguments in the Senate and on the political hustings have imbued my views with excessive caution. But I believe that we must wrestle successfully with our penchant for protectionism, our worries over the mercantilist aggression of some of our friends, immigration, support for democratic institution building, and occasional traces of racial prejudice if we are to enjoy another decade of security in the Pacific. Ultimately, there will be no substitute for wisdom and sound decisions.

The United States has pursued four fundamental goals through our foreign and security policy in the post–World War II period. They include the goals of defending and advancing the cause of democracy, freedom, and human rights throughout the world; of promoting prosperity and social progress through a free, open, and expanding market-oriented global economy; of working diplomatically to help resolve dangerous regional conflicts; and of striving to reduce and perhaps eventually eliminate the danger of nuclear war. We have sought with increasing urgency over the past two decades to meet our responsibilities and achieve those goals in the Asian-Pacific region. There and elsewhere, we have acted in the belief that America's prosperous and peaceful future can only be assured in a world in which other people, too, can determine their own destiny, free of coercion or tyranny.

The prospects for such a future in the Asian-Pacific region seem brighter than at any previous time in history. Yet Americans should not blind themselves to the obstacles that may obstruct such a future.

The United States cannot meet its responsibilities or protect its interest in the Asian-Pacific region without an active yet prudent diplomacy buttressed by American economic and military power. The United States cannot be expected to devise solutions to insoluble problems in the Asian Pacific. But neither should it proceed in a half-hearted manner when there are prospects of success in the area. Wishful thinking reflective of peculiar ethnocentric tendencies and vacillating commitments born of economic pique will not protect or promote the interests of the United States in the Asian-Pacific region.

Only those leaders at home and abroad who are able to rise above such sentiments stand a reasonable chance of confronting in a constructive fashion both the challenges and opportunities— be they in the areas of balance-of-payments problems, currency realignments, base negotiations, burden-sharing, cultural diversity, or human rights—that will manifest themselves in the Asian-Pacific region during the next decade and the coming century.

ONE

Regional Security
in Asia and the Pacific

June Teufel Dreyer

Introduction

That the twenty-first century will be the "century of Asia" has
by now become an accepted—if still unproven—opinion in
many parts of the Western world. Certainly the Asia-Pacific
region has enjoyed spectacular economic growth in recent years, its
dynamism presenting a sharp contrast to the disappointing economic performances of Africa, Latin America, and the Middle East.
In the mid-1970s, the United States' trade with Asia and the Pacific
surpassed that with Europe for the first time. It has continued to do
so each year since then, with the exception of 1979, and by ever
larger amounts. In 1987, the figures were \$240.8 billion for Asia and
the Pacific, vis-à-vis \$169.6 billion for Europe. In other words, trade
with Asia and the Pacific was 42 percent higher.[1]

There has been intense discussion within the United States as to
how it should respond to the Asian challenge. Most of the debate
has concerned the economic aspect of America's relationship with
the Eastern Hemisphere: should the United States practice protectionism, should it press other nations to open their markets to
American goods, should it concentrate on developing and more
aggressively marketing higher-quality products, or should it strive
for some judicious balance of all of the above? Relatively little
attention has been given to the defense aspect of America's position in Asia and the Pacific.

Karl Marx believed that changes in ideology lag behind
changes in the relationships of production. *Mutatis mutandis,* the
same might be said to apply to American strategic thought. The
United States, while acknowledging the increasing importance of

its ties with Asia and the Pacific, has continued with its military posture of preparing for one and a half wars, at least one of which is assumed to be in Europe. About 75 percent of all land-based forces stationed abroad are in Europe, vis-à-vis 20 percent in the Pacific.[2] Yet, in the opinion of most knowledgeable observers, a war in Europe appears increasingly less likely. Given the even lesser likelihood of a confrontation elsewhere, it would seem prudent to discuss how scarce resources might be deployed to meet the eventualities.

To the extent that debate has taken place on these issues, the choices have not been easy ones. In contrast to deep and long-standing ties with the European continent dating from the exploration of America, knowledge and depth of understanding of Asia have been scant. Few of America's citizens have claimed Asian-Pacific ancestry. The number, though growing, is still comparatively small. The region is not only less familiar culturally, it is of much greater size and presents a much greater diversity of ethnic groups, cultures, and religions than the relatively homogeneous white Judeo-Christian cultures of Europe. While New Zealand and Australia are basically white and Judeo-Christian and fit well into the European cultural context, Malaysia, Brunei, and Indonesia are overwhelmingly Muslim, and there are important Muslim minorities in the Philippines and Thailand. Hispanic Catholicism is the religion of the great majority of Filipinos, who are ethically similar to the Malays. The cultures of Thailand, Laos, and Cambodia are heavily influenced by Indic civilization. Those of China, Taiwan, Japan, the two Koreas, and Vietnam are Confucian with significant Buddhist influences, and in the case of Japan, Shintoist as well. In the South Pacific, Christianity is dominant, with modifications according to the respective Melanesian, Micronesian, and Polynesian cultures of its practitioners.

Though United States involvement in the Asian-Pacific area is often attributed to the outcome of World War II,[3] the American presence there was in fact substantial for more than a century preceding the war. American missionaries were active in most areas of Asia and the Pacific, and more than a few Americans developed a fascination with the languages and cultures of various areas within the region. From the earliest days of the American republic, the United States had important whaling activities in the Pacific. Oil from whales, many of them caught in the Pacific, was traded to China for tea and other Chinese products. The East Asiatic Squadron of the U.S. Navy, based at Hong Kong from the time of the Opium War until 1898, was a major component of the fleet. The Open Door Policy, promulgated in 1900, was intended as much to further already extant American commercial interests in China as for reasons of altruism. When Commodore Perry visited Japan in the mid-nineteenth century, his original mission was in consequence of these long-standing American Pacific interests. As an index of the importance Perry's government placed on his mission, the Black Ships of his squadron constituted approximately one-fourth of the United States fleet of the time. After the Philippines became an American colony in 1898, and up until 1941, the defense of the Philippines was the U.S. Navy's most important mission.

It should also not be forgotten that the Japanese decision to go to war with the United States in 1941 was a direct result of America's total embargo on sales of oil and other items to Japan. That embargo was in turn a response to Japan's extension of her control in Indochina to include South Vietnam. The United States might arguably have escaped involvement in World War II had it not been for American commercial interests in Asia.

The outcome of the war in the Pacific continued, on an enlarged scale, America's participation in the politics of the area. The experiences of "flying the hump" over the Himalayas into China, island-hopping across the Pacific, military cooperation with Australia, and defeating and occupying Japan led to an enhanced American awareness of Asia. World War II enlarged the military dimension of American interests that had heretofore been primarily commercial and cultural. This trend continued with the advent of the cold war, including a conflict in Korea that drew United States troops into battle with the Chinese military. American strategy at this time was premised on preparation for two and a half wars, with China and the Soviet Union as primary adversaries.

The Sino-Soviet dispute of the 1960s and subsequent Chinese moves toward normalization with the United States cleared the way for Sino-American rapprochement. This rapprochement occurred just at the time that rapid economic progress began to take place in Asia. The sharp jump in oil prices during 1973–1974 resulted in a worldwide recession and a debt crisis for less-developed countries (LDCs). States recovered differentially from these so-called "oil shocks," with the countries of East and Southeast Asia proving more adept at coping than most others. Their basically externally oriented policies encompassed a range of different economic strategies. At one extreme, Singapore and Hong Kong have relied on free trade practices for their success. Taiwan and South Korea followed policies of liberalizing exports as one ingredient of export-led growth strategy. ASEAN countries maintain different degrees of protection for their manufacturing industries, with Malaysia's being moderate, and Indonesia's more restrictive. However, the degree of protectionism in ASEAN as a whole is comparatively less than in LDCs elsewhere who have similar economic structures.

The pattern of this developed and expanded trade has been concentrated mainly within the Pacific Basin. The newly industrializing countries (NICs) specialize in manufactured goods and sell between 50 and 70 percent of their exports within the Pacific Basin. The United States is by far their biggest market. The other Asian LDCs export primarily natural resources and sell from 50 to 80 percent of these exports within the Pacific Basin.[4] Japan is their largest market, as it is also for the primarily raw-material exports of Australia and New Zealand. The People's Republic of China is Australia's fourth largest export market,[5] and New Zealand's fifth largest.[6] Japan has also been the major market for Chinese exports, after Hong Kong, and the United States is third. The import–export markets of the many small South Pacific states are also within the Pacific Basin, looking mainly to Japan, Taiwan, Australia, and New Zealand for both trade and aid.[7]

The United States remained on generally friendly terms with the states of Asia and the Pacific, with the exception of Vietnam and North Korea, during the 1970s and 1980s. For a time, however, the Soviet Union appeared to be assuming an ever more menacing posture. By 1988, there were fifty-seven Soviet tank and motorized infantry divisions in Asia, up from forty-three a decade earlier. The Soviet strategic air force includes over eight hundred warplanes, including some of the most advanced Soviet models. In 1983, the Soviet Pacific Fleet became, and remains today, the Soviet Union's largest. It is second only to the Northern Fleet, which opposes NATO, in strategic striking power. The Pacific Fleet continues to expand, as demonstrated by the recent addition of five new guided missile destroyers, and receives newly constructed major combatant ships, submarines, and auxiliary vessels at a pace nearly equal to that of the Northern Fleet.[8] The Pacific Fleet's reach has been extended by the acquisition of basing rights at Cam Ranh Bay in Vietnam. The facility there is the largest Soviet naval base outside the Soviet Union; typically, it comprises about twenty-five ships, two to four submarines, and nearly forty reconnaissance, ASW, strike, and fighter aircraft.[9] In the latter half of the 1980s, the Soviets signed successive fishing treaties (since lapsed) with the small South Pacific states of Kiribati and Vanuatu. Because each involved sums of money considerably in excess of the market value of the fish, there was concern that strategic rather than commercial interests had been the motivating factor.

Policy Options

How to respond to this perceived challenge to its Asian interests became an increasing concern of United States policy makers, particularly after the Soviet invasion of Afghanistan in 1979. Various schemes were discussed, including a Sino-American strategic alliance, a Sino-Japanese alliance, and an alliance of all the noncommunist states of Asia.

The idea of a Sino-American alliance was in fashion during the Carter years and in the early years of the Reagan administration. Soon after Richard Nixon's visit to the People's Republic of China in 1972, the two-and-a-half-wars strategy became a one-and-a-half-wars strategy, with China no longer viewed as a hostile country. After the Soviet invasion of Afghanistan, the Carter administration announced that America's previous policy of evenhanded treatment of the Soviet Union and China in terms of technology transfer would be amended in order to favor China. The United States also agreed to consider the transfer of military-related and dual civilian-military use equipment to the People's Republic. The concept of an anti-Soviet alliance based on an amalgam of Chinese manpower and American technology was attractive to several key policy makers. Counterarguments were raised by China specialists who pointed out that

1. Rather than being intimidated by such an alliance, the Soviet Union might be provoked to respond with hostility;

2. Providing the military technology for China to stage a credible defense against the Soviets would cost an unacceptably high figure, in excess of several billion dollars;

3. Educational levels in the Chinese military are so low that soldiers could not make proper use of advanced American technology in any case;

4. As shown by the People's Liberation Army (PLA)'s performance in its invasion of Vietnam in 1979, the Chinese military was unable to project power any distance beyond its borders; and

5. The People's Republic could not be counted on to side with the United States in the event of a Soviet-American conflict.

Eventually, China's own cool reception to the idea of a strategic alliance, and its moves toward reconciliation with the Soviet Union in 1982, spelled the demise of this concept.

A Sino-Japanese alliance struck some American policy makers as desirable since it meant the Asianization of the defense of Asia and the Pacific. They argued that the cultures of the two countries were similar and that thus there was an underlying basis of mutual understanding and cooperation. Moreover, the economies were complementary, with China having an abundance of natural resources and a shortage of capital, technology, and managerial expertise, while Japan had a shortage of natural resources but an abundance of capital, technology, and managerial expertise.

That the cultures of China and Japan are similar is true in the broad sense. But the tacit assumption that there is more than a broad similarity could only be made by those who have never carefully studied the cultures of the two countries. Moreover, the assertion that even very similar cultures necessarily foster understanding and cooperation runs counter to much of Western as well as Oriental historical experience. The case of northern Ireland is only one of numerous examples that could be cited. Alleged similarities of culture did not prevent China and Japan from fighting two exceedingly bitter wars in recent times. That harsh memories survive was shown by China's very sharp reaction in 1982 to news that the Japanese Ministry of Education planned to revise the nation's textbooks in a manner that would describe Japan's conduct during World War II in a more favorable way. And arguments based on the complementarity of economies ignore the possibility that Japan might find more favorable opportunities for investment of its capital elsewhere. They also fail to take into account the existence of a Chinese internal market that is so large as to limit the amount of natural resources China has available to sell abroad.

At present, Sino-Japanese relations are diplomatically correct and trade relations are recuperating after the Japanese government suspended certain loans in the wake of the Tiananmen incident. However, the relationship has been accompanied by considerable strain and mistrust. Moreover, the two countries do not always see the Asian diplomatic scene in the same light. For example, Japan has made important investments in Vietnam, with whom China has been intermittently at war since 1979. The accumulated weight of these factors means that it is highly unlikely that the peace and security of the Asian-Pacific region can be ensured by a strategic alliance between Japan and China.

A third scheme that has been advanced for ensuring peace and stability in Asia and the Pacific involved an alliance of all the noncommunist states of the area. While this idea had the advantage of ideological neatness, it fell athwart the problem that not all of the noncommunist states of the area had the same perception of threat. The United States and Japan clearly viewed the Soviet Union as their major adversary. South Korea worried principally about North Korea; the Republic of China on Taiwan worried principally about the People's Republic of China on the mainland. Malaysia and Indonesia worried about subversion directed from the mainland as well. Thailand saw Vietnam as its short-term threat and the People's Republic as its longer-term adversary. Brunei was mainly concerned with Malaysia, and the Philippines and Burma with internal insurgencies. A number of Australians worried about being inundated by an influx of Indonesians, and many New Zealanders were convinced that there was no threat at all. Implementation of the noncommunist alliance would also necessitate solving some other bothersome problems, such as what would happen to America's budding relationship with the People's Republic should the United States include the Republic of China in an alliance system. It also proved erroneous to assume that all noncommunist countries would be willing to cooperate with each other: South Korea, for example, has steadfastly refused to cooperate with Japan, even on so obviously desirable a matter as the security of the strait that separates them (and which, symbolically, they refuse to call by the same name). There was also the concern that lumping all the communist states together in the adversary category would in fact drive them closer together. A revival of the Sino-Soviet alliance was not something the United States wished to abet. Eventually, this scheme also fell of its own contradictions.

The advent of *glasnost* and *perestroika* in the Soviet Union introduced new complexities into defense planning for Asia and the Pacific. While most analysts believed that Gorbachev was sincere in wanting to cut back military expenditures and assume a less threatening posture, many doubted that his reforms would succeed. There were also concerns that the strains these reforms generated would force Gorbachev from office, and that he would be replaced by a more conservative government that would resume the aggressive policies from which Gorbachev had sought to distance himself.

Further uncertainties were raised by China's repudiation of large segments of its reform program and brutal repression of citizen protest movements in the spring of 1989. The Sino-Soviet summit of May 1989 was all but obscured by the demonstrations taking place in Tiananmen Square, just outside the summiteers' meeting place. The new, more conservative Chinese leadership has been quietly critical of Gorbachev's policies; an internal party memorandum has accused him of "subverting socialism" and held the Soviet leader responsible for the turmoil in Eastern Europe.[10] At the same time, the People's Republic began moving closer to other communist states that have resisted liberalization, including Cuba and North Korea. How much of a realignment will actually take place and whether it will necessitate a major rethinking of United States strategy in Asia and the Pacific remain to be seen.

Lacking more elegant conceptual alternatives, the United States has sought to maintain a balance of power in Asia and the Pacific through a variety of bilateral and multilateral ties. Some of these are formal, and the United States participates in them directly. These include mutual cooperation and security treaties with Japan, South Korea, and the Philippines, plus a commitment to Thailand based on the Manila Treaty of 1954. The United States was also a signatory to the tripartite ANZUS treaty, with Australia and New Zealand, though ANZUS now seems to survive as two separate relationships, one between the United States and Australia and a second between Australia and New Zealand. Theoretically, the United States has no official relationship with the Republic of China, though it continues to supply Taiwan with arms and technology under the terms of the Taiwan Relations Act of 1979.

Other United States participation is indirect. America has supported the Association of Southeast Asian Nations (ASEAN). The organization publicly disavows any collective military role, partly for fear of antagonizing Vietnam and partly because the other members fear domination by Indonesia, which has by far the largest military among the membership. Bilateral military cooperation is, however, encouraged.[11] The most recent example of this is the inauguration of a joint air weapon range for the armed forces of Singapore and Indonesia.[12] Also, most ASEAN members have received military aid from the United States and/or Great Britain. The Five Power Defense Arrangement, including Australia, New Zealand, Great Britain, Malaysia, and Singapore, aids Malaysia and Singapore to develop their defenses, and the Integrated Air Defense System provides these two ASEAN members with a mechanism for coordinated air defense. Singapore also trains its troops in Taiwan under an arrangement called "Starlight Operations"[13] that exists despite the absence of formal diplomatic relations between the two.

Military cooperation between the United States and the People's Republic of China has proceeded slowly, since both sides have reason for caution. Neither wishes to provoke the Soviet Union. The issue of continuing American arms sales to Taiwan also remains sensitive. Since the Taiwan Relations Act commits the United States to maintaining a military balance in the Taiwan Straits, some American policy makers have reasoned that sales of military equipment to the People's Republic should be balanced off by equivalent sales to the Republic of China. The People's Republic has objected strenuously. Nevertheless, during the past three years, the United States has provided China with advanced avionics for its F-8 high-altitude interceptor. There has been a passing exercise involving Chinese and American ships, and United States Navy ships have made a port call at Qingdao. China has reciprocated by sending a training ship to Honolulu.[14] A small exchange program exists between the respective, and identically named, National Defense Universities of the two countries. Just after the Tiananmen incident, responding to pressures from

Congress, President Bush imposed an embargo on military sales to the People's Republic. However, this was clearly meant as a temporary measure rather than a termination of the cooperative relationship.

At the same time, the United States has transferred technology to Taiwan for the development of its indigenous defense fighter (IDF),[15] and cooperated in the construction of frigates for its navy.[16] The country retains its direct link into the Pentagon computer for the purchase of spare parts.

The above measures, combined with renewed and reasonably successful efforts to induce Japan to play a larger role in the defense of the region, would seem to represent what is realistically possible in terms of Asian-Pacific security. Recently, a new set of voices has begun to be heard, not arguing as had others in the past, that these efforts are not enough. To the contrary, these critics reason that America's efforts on behalf of regional security are in fact more than is needed, and as such are both unnecessarily expensive and potentially provocative.

As expressed by a distinguished senior staff member of the Bangkok daily *The Nation,* the outbreak of peace in Asia and the Pacific has rendered the United States forward force deployment strategy obsolete.[17] A good deal of credit for this belongs to Soviet leader Mikhail Gorbachev. The pledges made in his July 1986 Vladivostok speech, to remove Soviet troops from Afghanistan and Mongolia,[18] have now been implemented. This fulfilled two of the three prerequisites China had stipulated before Chinese-Soviet relations could be normalized. Gorbachev also announced that he would accept Beijing's claim that the main channel rather than the Chinese bank should be used for purposes of demarcating the two countries' disputed river border.

China's third condition, withdrawal of Vietnamese troops from Cambodia, has also been effected. The People's Republic complains loudly that many soldiers remain, on pretexts such as their being advisers to various projects or because they are married to Cambodian women. However, the complaints appear to be motivated more by a desire to increase China's diplomatic leverage and to help the long-term Chinese clients, the Khmer Rouge, than by the sincere belief that the Vietnamese have made only a sham withdrawal.

A second speech delivered by Gorbachev, at Krasnoyarsk in

September 1988, was entitled "Strengthening Security in the Asian-Pacific Region." In it, Gorbachev advanced seven proposals, including pledges that the Soviet Union would not increase its nuclear stockpile in the region, that it would consult with other powers with regard to limiting naval and air force activity, and that it would withdraw from Cam Ranh Bay provided that the United States would relinquish its bases in the Philippines.[19] The speech received a less than enthusiastic reception, with American sources pointing out that, given the long distances between the United States and most Asian states, vis-à-vis the Soviet Union's relative proximity, the value of Cam Ranh Bay to the Soviets was much less than that of Clark and Subic Bay to the United States. It is worth pointing out, however, that although the Tashkent speech also received an initially cool reception, it eventually produced tangible results. Gorbachev's Krasnoyarsk speech may do so as well.

Although the Sino-Soviet summit of May 1989 did not solve any important issues or set new directions for future relations, it did constitute a formal end to thirty years of conflict.[20] Whether the different policies pursued by the two states since that summit will cause them to drift apart remains to be seen. For the moment, formal relations are cordial and trade continues to grow.

United States–Soviet hostilities abated also, as symbolized by the successful conclusion of the Intermediate Nuclear Forces (INF) negotiations in the closing days of 1987. Progress was made in other areas of arms control as well, prompting the publication of an article in a recent *Proceedings* of the U.S. Naval Institute entitled "Our Peaceful Navy: Who's Left to Fight after Our Main Adversary Walks Away?"[21]

Several other traditional adversaries began to mend relations at the same time. In 1987, the Taipei government announced that it would no longer object to its citizens' visiting the People's Republic for reunions with family members. Many thousands of persons have made the trip, allegedly to see long-lost relatives but often out of simple curiosity or for business purposes. The mainland government frankly encourages investment from Taiwan; the Taiwan government continues to officially forbid direct trade, while merely warning business people to avoid becoming too dependent on mainland markets. Indirect trade has soared in recent years. The exchange has been good for both sides: the mainland economy has been helped by Taiwan capital and managerial expertise, and the Taiwan government has had its public image bolstered for having taken the initiative. In addition, the overwhelming number of Taiwan citizens who have transited the Taiwan Straits have had indelibly imprinted on their collective consciousness the enormous disparity between their relatively affluent living standards and the appalling poverty on the mainland. This, too, has undoubtedly helped the Kuomintang government's image with the electorate. Tensions between the People's Republic and the Republic of China government seem to be at an all-time low. According to mainland statistics, 450,000 people from Taiwan visited the PRC in 1988, indirect trade reached $2.5 billion, or 65 percent more than in 1987, and 3.4 million letters were exchanged.[22]

The Tiananmen incident had little overt effect on the growing relationship between the two Chinas. Taipei condemned the violence and enacted measures to help dissident intellectuals who had escaped from the mainland. But investors from Taiwan were among the first to return to the mainland. Surely more wary, and increasingly concerned with the effects of political instability in the People's Republic, they nonetheless returned. China's austerity program rather than renewed hostility between the two countries seems to be the reason for a slowdown in trade across the Taiwan Straits.

The two Chinas also seemed to be making progress in sorting out their respective diplomatic relations with other countries. In early 1989, the People's Republic and Indonesia began to negotiate to reestablish formal ties, which had been broken during the mid-1960s. Singapore announced that it would recognize the People's Republic after Indonesia did so. At the same time, both countries made it clear that their trade and other ties to Taiwan would be maintained. In fact, Republic of China president Lee Teng-hui was being received with full honors in Singapore when the intention of reestablishing ties with China was announced, and all sides pronounced his visit a great success.[23]

Similarly, the reaction of the People's Republic to the Republic of China's announcement of formal diplomatic ties with the Bahamas was low-key. The People's Republic has repeatedly denounced Taiwan's new "elastic diplomacy," under which the Taipei government decided to re-enter international organizations and reestablish ties with countries with whom it had severed relations when those entities recognized the mainland government. The People's

Republic then broke relations with three countries that decided to establish formal ties with the Republic of China but has made few other noticeable efforts to counteract elastic diplomacy.

There are even indications that China's relations with its most recent worst enemy, Vietnam, may be improving slightly. The level of vitriol in each side's media discussion of the other has dropped markedly, and trade relations have improved concomitantly.[24]

South Korea also began to trade with China in 1987, overcoming its long-standing opposition to dealing with communist regimes. There is an excellent market for China's coal in South Korea and Korea's high-quality consumer goods have been well received in the People's Republic. Despite the absence of formal diplomatic relations between the two (South Korea recognizes the Republic of China, while the People's Republic recognizes North Korea) direct shipping relations between Pusan and Inchon in South Korea and Shanghai and Tianjin in China began in May 1989. The vessels will fly third-country flags.[25]

South Korea's superb management of the 1988 Summer Olympics, despite the threat of North Korean terrorism and the existence of considerable domestic political discontent, facilitated the government's efforts to expand trade relations elsewhere in the communist world. Within months after the close of the games, trade offices had been exchanged with Hungary, Yugoslavia, and the Soviet Union.[26]

Subregional Considerations

China

How to tailor policy to a region undergoing such rapid economic, social, and security changes was the subject of a seminar series sponsored by the Washington Institute in 1988–89. In his chapter on China, Donald Zagoria characterizes the People's Republic as maintaining a low-posture security policy designed to buy time for modernization. During this period, and given its modest military capabilities, China's best strategy is to play off the United States and the Soviet Union against each other. The People's Republic has indeed maneuvered itself into the swing position between the superpowers and was able to manage balance-of-power politics with skill for some time. However, given the recent lessening of tensions between the United States and the Soviet Union, this role may become increasingly difficult for China to play.

Sino-Soviet relations will continue to be constrained by a lack of mutual trust, by geopolitical competition in Asia, and by the imbalance in the two countries' military strength. Moreover, the different policies that China and the Soviet Union have been following since Tiananmen may contribute to these frictions.

At the same time, Sino-American relations have deteriorated since the Tiananmen incident. The People's Republic accused the United States of having instigated the protest movement and of seeking to export bourgeois democratic values to its territory. Specific events, such as the decision of leading dissidents Fang Lizhi and his wife to take refuge in the United States embassy in Beijing, added to the strains.

Problems in Sino-Soviet and Sino-American relations, when added to the lessening of tensions between the United States and the Soviet Union, point to reduced Chinese leverage within the triangular relationship. Moreover, although Zagoria does not anticipate any dramatic changes in Chinese foreign policy as a result of the Tiananmen crackdown, he notes that the People's Republic has lost a substantial amount of international good will as well as economic aid.

For at least the past decade, Chinese foreign policy has been characterized by four priorities:

1. A desire for integration with the world economy so as to support China's ambitious plans for economic development;
2. The perceived need for a peaceful international environment, particularly in Asia, so that the People's Republic can minimize its military expenditures and devote maximum effort to domestic affairs;
3. A desire for stable relationships with both superpowers, while avoiding confrontation or alignment with either;
4. Progress toward reunification with Hong Kong and Taiwan.

Its repression of the pro-democracy movement notwithstanding, there has been nothing to suggest that Beijing has lost interest in any of these priorities, or that it is planning to return to the revolutionary foreign policy it conducted during much of the Mao era. However, the worsening economic situation and domestic instability that followed the suppression of the movement may have made the attainment of these objectives more difficult.

Uncertainties about China and its future development will continue. At the same time, there are concerns about the course of events in Eastern Europe and their effect on both the global and regional balances of power. Despite much talk about the "decline" of the United States, Zagoria notes that every one of America's allies in Europe and Asia wants to see a continuing American military presence on the Eurasian continent. Even the Soviet Union may eventually come to realize that a strong American military presence in the Pacific is stabilizing and possibly even preferable to some of the alternatives. And, finally, for most Asian countries there is simply no substitute for the American market.

Japan

Donald Hellman's chapter on United States–Japan security relations judges the alliance between these two states to be one of the most successful and long-standing achievements of postwar American foreign policy. However, due to the ongoing and far-reaching structural changes in the global and Asian-Pacific regional economies, he is doubtful that it can continue in its present form. The more pluralistic world created by the post–World War II Pax Americana makes change from hegemony to partnership an international imperative for policy makers in Tokyo and Washington.

Hellman describes Japan as more like a trading company than a nation-state, and as without a true foreign policy. Yet, as a global economic superpower with potential for, if not necessarily aspirations to, world leadership, Japan should not be expected to remain in a protectorate status derived from its "peace" constitution, written in a week by the United States occupation forces. At present, there is a gap between Japan's huge economic power and any defense commitments beyond token delaying actions for its own territorial defense. Hellman suggests a number of ways in which this gap might be filled, ranging from increased burden-sharing to integrated defense planning and multilateral force structures. He rejects as bizarre the recent proposal by then-Prime Minister Noburo Takeshita to substantially expand Japan's aid to the third world in lieu of expanding his country's security role. For the United States to agree to this would be tantamount to its accepting a division of labor in which America plays the role of international policeman while Japan becomes international Santa Claus, with the added benefit that roughly 70 percent of Japan's foreign aid is tied—in fact, if not formally—to the Japanese economy.

If the United States and Japan are to successfully adjust their bilateral relationship, both countries will have to undertake substantial domestic institutional reforms that are tantamount to *perestroika*. In large part due to the informal, non-legal web linking Japan's ruling political elite to its society, a Japanese initiative in foreign policy is far less likely than an initiative from America. The United States, Hellman concludes, has not only the capacity but the obligation to fashion a Pax Americana with Japan a full partner.

Korea

The thesis of Edward Olsen's chapter on contemporary United States–Korean security relations is that the growth in South Korean economic power and military competence has made the United States presence there increasingly unnecessary. As a consequence of having worked itself out of a job, the United States should be increasingly concerned with solving what problems remain, while transferring as much as possible of the responsibility and cost of South Korean security to the Republic of Korea itself. Alternatively, it should encourage closer US–ROK cooperation on regional security issues.

Meanwhile, there are numerous minor and four major areas of disagreement between the United States and Korea. Of the major issues, Olsen believes that the easiest of the four to solve is the nuclear question. Although America has a long-standing policy of neither confirming nor denying the existence of nuclear weapons anywhere, the media in the United States and in both North and South Korea generally assume that nuclear weapons are in fact present in Korea. A certain amount of tension is caused thereby, with some Koreans worried that the United States may prefer to use these weapons in Korea rather than against Warsaw Pact enemies. Paradoxically, other Koreans are concerned that the United States lacks the will to use nuclear weapons at all, even should the defense of their country necessitate such a decision. Although these disparate fears raise uncertainties within the security relationship, the nuclear aspect of bilateral tensions is muted by each country's reluctance to speak openly on the topic.

The second major problem area concerns the respective roles of the United States and the Republic of Korea in the Combined Forces Command (CFC). Olsen judges that each side is guilty of arrogance toward the other. American arrogance derives from great-power chauvinism; Korean arrogance, from a complex blend of confidence in the superiority of Korean martial values, pride in Korea's ancient culture and homogeneous ethnic background, and reaction against American arrogance. The fact that an American general is head of the CFC is extremely annoying to many Koreans.

A third serious area of disagreement is some Koreans' belief that, by manipulating the South Korean military, the United States is perpetuating a puppet dictatorship in their country. Because of American dominance of the CFC, denials tend to be unpersuasive. A corollary of this, the rise of anti-American sentiment because some Koreans think that the United States is standing in the way of the Republic of Korea's sometimes disorganized and often disrupted course toward democracy, is also a serious problem.

The fourth, and most sensitive, area concerns the potential role of Japan in Northeast Asian security, and more specifically, in Korean security. Koreans' intense dislike of Japan has caused the United States to maintain an artificial barrier between its security policies toward each. As Olsen points out, the threats to both countries emanate from the same source, the Soviet Union, and the geopolitical significance of each state to the United States has been used to justify American commitments to the other.

The first three of these issues will presumably disappear after the United States withdrawal from Korea; the fourth will almost certainly become more serious. Meanwhile, all will have to be treated with utmost care so as to avoid repercussions for the security of the peninsula.

Taiwan

Harry Gelber characterizes the Republic of China on Taiwan as being in a period of transition. With its economy booming and its polity in the midst of an impressive transition to democracy, the island state recently began to redefine its international status through the medium of elastic diplomacy.

Taiwan's security concerns, Gelber notes, come in a variety of shapes and sizes, with the overwhelming fear being an attack from the mainland. Although currently this seems neither likely nor even feasible, the major thrust of the Republic of China's defense planning must continue to be making such an attack as difficult and costly as possible for the People's Republic. In addition to having enough of the right weapons, this entails several ancillary concerns: Taiwan must watch the political and military consequences of the Sino-Soviet dispute and shore up its relationships with as many other countries as possible.

Gelber points out that Taiwan has considerable strategic importance for the major powers involved in Northeast Asia. The island lies astride the Taiwan Straits, commands the southern access routes to Japan, and dominates the Eastern Sea approaches to the People's Republic. It is an important link between the Pacific and Indian Oceans. Its loss would break the line of islands, all of them friendly to the West, that lie off the East Asian mainland and that have since World War II been regarded as of great importance for the West's strategic posture in the Pacific region and beyond.

While the People's Republic and the Republic of China have discussed various interpretations of the "one country, two systems" scheme for reunification, the possibilities for effecting it seem remote at present. China's brutal suppression of its own citizens' protest movement and its harder line toward Hong Kong have not inspired confidence among the citizenry of the Republic of China. Moreover, the increasing Taiwanization of the Taipei government means the coming to power of a group that has little interest in reunification. For the moment, Gelber advises, the Beijing regime would be wise to let relationships with Taiwan grow naturally at people-to-people and business levels, in ways that will be of fairly direct help both to ideas about "Great China" and to the economic refurbishing of the mainland.

The United States continues to have a number of interests at stake in regard to Taiwan. Washington must see that the Republic of China does not conclude other alliances that are potentially harmful to the United States. Gelber suggests that, should the United States be denied access to bases currently located in the Philippines, it might wish to consider emplacing them on Taiwan. There are, he points out, major difficulties in doing so. The foremost among these would be Beijing's adverse reaction. Also, precisely because Taiwan is of such strategic importance, it is probably in the interests of all major Pacific entities that the island should be independent (de facto if not de jure), equipped defensively, and not able to act as a base for any of the major powers.

Indochina

Douglas Pike argues that Vietnam's chief threat to American interests derives from its association with the Soviet Union. Vietnam does not present a credible military threat to any American associate or friend in Southeast Asia except Thailand, since it does not have the air or naval power to project forces over great distances. The Vietnamese military could invade and occupy Thailand, but at the risk of extending its problems with Cambodia to noncommunist Southeast Asia. It could probably hold its own in a limited war with China but could not sustain this effort indefinitely.

A second threat from Vietnam is the indirect one of its funding and supporting insurgencies in ASEAN countries. These could prove troublesome and costly to suppress. Persistent but unconfirmed rumors in 1988 alleged that Vietnamese weapons and military supplies were reaching the communist New People's Army in Luzon and the Moro insurrectionists in Mindanao. Pike believes, however, that Hanoi has ruled out aid to insurgencies for the present.

A third threat from Vietnam might be labelled politico-institutional, occupying the intermediate area between insurgent war and politics. In the flush of its victory of 1975, Hanoi mapped a strategy for the Southeast Asian region that called for an end to economic cooperation with Western and Japanese capitalism. It hoped to shut the United States out of the region entirely through this plan. Since Vietnam's time of troubles began in 1979, little has been said publicly about the scheme. But the idea is still alive in Hanoi, and may well be resuscitated.

At present, Vietnam's official attitude toward the United States is that it wishes to forgive and forget. Pike argues that this attitude should be evaluated in the context of Vietnam's historical record and tradition of seldom forgiving and never forgetting. There is an intense political struggle now taking place within the Vietnamese leadership, with no indication that it will soon be resolved. Domestic problems, particularly economic ones, preoccupy the elite group, and the matter of future relations with the United States is a secondary consideration.

It is probable that, at some time during the Bush administration, formal diplomatic relations will be established with Hanoi, at either the interest-section level or the embassy level. However, Pike counsels, recognition will not automatically bring benefit to the United States. There is no necessary cause-effect relationship between recognition and a diminution of Soviet influence in Indochina, reduction of Vietnam's influence in Cambodia or Laos, enhanced United States business activities in the area, or better access to Hanoi's prisoner of war files. This is not, Pike stresses, an argument against recognition, but simply to point out that recognition is one thing and problem resolution is another.

Pike concludes that America's future relationships with Vietnam and the rest of Southeast Asia will not be dominated by national security considerations as in the past, but will increasingly involve economic factors. To date, however, the fact that Vietnam is desperately poor and needs much foreign assistance has not caused the country's Politburo to behave in an economically rational manner, nor has it significantly influenced either foreign or domestic policies.

The Philippines

Lawrence Grinter's chapter on United States–Philippine security relations points out that, alone among America's Asian-Pacific allies, the Philippines insists on United States military and economic aid in return for being defended by the United States and hosting the American forces that defend it and help underwrite the strategic balance in East Asia and the Pacific. In the lengthy and relatively close relationship between the two countries, the Filipinos have basically wanted respect and the Americans have wanted appreciation. Neither has gotten enough of what it sought.

Every post–World War II Philippine president has talked about ending the "special relationship" between the two countries, which Filipinos characterize as unequal and patronizing. However, fear of loss of the economic and other benefits of ties to the United States has dissuaded them from actually taking this step. The Aquino government administers a severely economically wounded democracy reflecting a number of crosscurrents and contradic-

tions. It has simply not come to terms with the security requirements of surviving independently in the changing Southeast Asian
context. The Philippine government is presently spending about 9
percent of its annual budget on defense. That is what is left after
social programs are funded and another 45 percent of the budget
goes to interest payments on the country's huge foreign debt.
President Aquino's position is that her country's geographic isolation allows it to be "indifferent" to external threats. Grinter specu-
lates that similarly comfortable views may have held sway in
1939–40, just before the Japanese invasion.

With the Military Bases Agreement between the United States
and the Philippines terminating in September 1991, just as the
campaign in the Philippine presidential election of summer 1992
is gaining momentum, there is a great deal of uncertainty about
the future of the United States–Philippine security relationship.
The options Grinter considers include

1. extending the current agreement,
2. ASEANizing the bases,
3. redesigning United States forces on a limited basis,
4. completely withdrawing United States forces and relocating
   the bases, and
5. gaining an understanding with Moscow.

Although public opinion polls show the majority of Filipinos
support an extension of the bases agreement, certain influential
politicians view the bases as little more than revenue-generating
devices that must be phased out before their country can regain
self-respect. Other leaders are more cynical, seeking to use the
bases issue as a tool for criticizing the government in order to
enhance their own prestige. Grinter advocates that the renegotiation process shift from a strictly bilateral to a multilateral framework in which more realistic assessments of interests are presented.

The Southwest Pacific

Owen Harries points out that after being considered of vital
strategic importance during World War II, the Southwest Pacific
became a geopolitical backwater due to a combination of American dominance and the Soviet Union's lack of strategic reach.
Recently, this has begun to change. Not only the two superpowers,

but other major countries, including France, China, and Japan, have begun to pay attention to, and formulate new policies for, the region. Even Libya and Cuba have been dabbling in the affairs of certain Southwest Pacific states.

While the economic dynamism of Asia and the Pacific has really centered on the Northwest Pacific and parts of Southeast Asia, concepts such as the Pacific Rim and the Pacific Community tend to be used inclusively, to encompass the South Pacific as well. Implicit in discussions of the dynamism and promise of the area is the belief that, if the hemisphere is destined to become the most important in the world, then even its more remote subregions will acquire new significance.

As the 1980s advanced, the internal politics of many of the newly independent states of the Southwest Pacific became characterized by corruption, instability, and violence. The island economies are poor, with few resources for development. Literacy is widespread, but there are few non-manual jobs to which education normally provides access. And a marked drift of population from countryside to towns is straining the urban infrastructure of the area. This is an explosive mix.

While the Southwest Pacific is rather marginal in terms of the overall Pacific naval scene, it would assume much greater importance should the straits of Southeast Asia be closed. Moreover, the three areas from which all Soviet space launches take place are located directly on the opposite side of the earth to a triangle in the South Pacific. Any satellite launched from these Soviet "space ports" will pass over this triangle. Control of the area therefore becomes important from the standpoint of surveillance.

For some years, the United States seemed to ignore the states of the Southwest Pacific. Foreign aid was minuscule, and the American Tuna Boat Association's penchant for fishing in what these states considered their territorial waters was an ongoing irritant. At the same time, the Soviets extended offers of fishing agreements, economic aid, trade development, collaborative oceanographic research, and visits to the Soviet Union. The Soviets also expressed support and sympathy for the islanders with regard to fishing violations, the struggle against French colonialism in New Caledonia, and the establishment of a South Pacific Nuclear Free Zone (SPNFZ). The Soviet Union enthusiastically ratified the protocols of the SPNFZ Treaty; the United States refused to sign. Other main targets of Soviet penetration have been labor unions, churches, and institutions of higher learning.

Harries argues that the redress of grievances that is required of the United States is neither particularly difficult nor particularly expensive. The signing of a fishing agreement in June 1988 was a step in the right direction. Attention and respect would do much to reduce the anti-Americanism that has grown in the Southwest Pacific in recent years. The islanders need to be conversed with more and lectured to less. More attention should be given to their economic concerns and less to the strategic threat. There is, Harries concludes, a paradox for the United States in this situation. The less seriously the region is taken, the more likely it is to become a serious security problem. Conversely, if it is treated as a region of some importance and attended to properly, it is likely to remain of modest significance.

Australia and New Zealand

William Tow's contribution deals with the consequences for Southwestern Pacific security of New Zealand's antinuclear stance and its rejection of NATO's "no confirm, no deny" nuclear policy on the entry of military vessels and aircraft. ANZUS, the loose but long-standing strategic alliance among Australia, New Zealand, and the United States, was transformed into a de facto grouping of two parallel alliances, with Australia as their common member.

New Zealand's challenge to United States alliance policies caused Washington to downgrade the value of that country's defense contributions. The United States made certain adjustments necessary to keep its South Pacific deterrence strategy intact. One was to restrict Wellington's special access to American intelligence and weapons systems. Another involved expanding American defense and economic assistance programs to South Pacific island states. Joint United States–Australian military exercises also became more extensive following New Zealand's exclusion from such maneuvers in 1985.

Tow argues that the United States and New Zealand, each in its own way, have been guilty of supplanting regional security interests with their respective agendas of extended deterrence and antinuclear politics. In doing so, they have unintentionally but unmistakably undercut the cohesion and viability of the Western defense that has served the region so well in the postwar era. Tow suggests that, as a first step toward rectifying this condition, the United States should ratify the Treaty of Rarotonga, with its establishment of a South Pacific Nuclear Free Zone. However, before such ratification, Washington would be justified in insisting that New Zealand compromise its antinuclear stance on a basis of "trust, but with no questions asked," as Japan has done.

In the absence of such a resolution, the three ANZUS signatories may still be successful in guiding the emerging South Pacific island states toward stability and economic progress. But the process will be more difficult. Moreover, the price for achieving a greater measure of security in the region could be even sharper divisions over what constitutes the best approach to global security.

Conclusions

At the beginning of decades and the start of new presidential administrations, it is common, and perhaps even clichéd, to make calls for "fresh new approaches," as indeed a distinguished scholar writing in a widely read journal recently recommended with regard to United States policy in Asia.[27] What these fresh new approaches might be was unfortunately not specified. Indeed it may be, as the author of Ecclesiastes noted several millennia ago, that there is nothing new under the sun. In the process of considering what the future course of United States policy should be for their respective subregions, chapter writers have advanced certain common themes.

First is the perception that in virtually every area of Asia and the Pacific, including the usually passive Burma, there has been rapid change. It is imperative that the United States monitor these changes carefully and tailor its policies to fit the new circumstances. Second, with regard to increasingly affluent and militarily competent allies, chapter writers generally favored burden-sharing arrangements with regard to defense. Allowing the ally to substitute increased economic aid to the region for participation in defense plans was not considered a good solution from the American point of view.

A third theme running through the chapters was the increasing importance of economic factors and disruptive population migrations, vis-à-vis more narrowly defined defense concerns, in ensuring the security of the region. There was also a consensus that bilateral arrangements and regional approaches be used to complement one another rather than as alternatives to one another. While the need to show respect and concern for other countries and their problems is little more than generality, chapter writers seemed in agreement that there is a balance to strive for. In showing this respect and concern, the United States must not back away when encountering resistance to its reasonable demands (a tendency in certain aspects of America's relations with both China and Japan). On the other side of the balance, the United States must seek to dispel suspicions that it pays too little attention to the sensitivities and points of view of local peoples, as many residents of the Philippines and the states of the Southwest Pacific believe.

The success of United States strategy in Asia and the Pacific is probably more dependent on America's ability to fine-tune an already basically viable strategy, or perhaps more accurately, a congeries of strategies, rather than on its ability to fashion a single grand strategy. Given the geographic extent, ethnic diversity, and economic complexity of the region, such overarching schemes are likely to prove unrealistic and ultimately harmful to American interests. Although, for the moment at least, great-power tensions are muted and most conflicts seem to be localized, the region does not lack problems with potential for spilling over into the international arena. Domestic violence in Burma and South Korea that is related to the respective countries' democratization efforts and popular unrest in China and the Philippines are the most salient recent examples. Though there are commonalities with other situations in other Asian states, each has its unique characteristics as well that must be recognized. As with liberty, eternal vigilance is the price of regional security.

NOTES

1. International Monetary Fund, *Direction of Trade Statistics Year-book* (Washington, D.C.: International Monetary Fund, 1988).
2. Leonard Sullivan, *Inside the 1988 Budget: Guidelines for American Fiscal Responsibility* (Washington, D.C.: The Atlantic Council, August 1988), p. 26.
3. See, for example, Toemsak Phalanuphap, "Peace Makes US Forward Deployment Strategy Obsolete," *The Nation* (Bangkok), 2 February 1989, in United States Department of Commerce, *Foreign Broadcast Information Service,* East Asia and the Pacific (hereafter, FBIS-EAS), 2 February 1989, p. 61.
4. This argument follows that made by Lawrence Kraus's "Regional Economic Trends," in Bruce Dickson and Harry Harding, eds., *Economic Relations in the Asian-Pacific Region* (Washington, D.C.: The Brookings Institution, 1987), p. 2.
5. *Xinhua* (hereafter, XH), Beijing, 13 November 1988, in United States Department of Commerce, *Foreign Broadcast Information Service,* China (hereafter, FBIS-CHI), 21 November 1988, p. 15.
6. XH in FBIS-CHI 12 November 1987; in FBIS-CHI 17 November 1987, p. 13.
7. See United States Central Intelligence Agency, *World Factbook 1989* (Washington, D.C.: Government Printing Office, 1989).
8. Frank Carlucci, *Annual Report to the Congress, Fiscal Year 1990* (Washington, D.C.: United States Department of Defense, 1989), p. 22.
9. *Soviet Military Power: An Assessment of the Threat 1988* (Washington, D.C.: United States Department of Defense, 1988), p. 124.
10. Reuters (Beijing), cited in "Worried Chinese Leadership Says Gorbachev Subverts Communism," *New York Times,* 28 December 1989, pp. 1, 5.
11. See, for example, *Bernama* (Kuala Lumpur), 23 March 1989, in FBIS-EAS 23 March 1989, p. 49; *Kyodo* (Tokyo), 24 March 1989, in FBIS-EAS 27 March 1989, p. 1.
12. *Antara* (Jakarta), 22 March 1989, in FBIS-EAS 23 March 1989, pp. 49–50.
13. *South China Morning Post* (Hong Kong), 17 February 1989, p. 10, in FBIS-CHI 23 February 1989, p. 71.

14. XH 31 March 1989, in FBIS-CHI 2 April 1989, pp. 5–6.

15. James McGregor, "Taiwan Unveils Home-Grown Fighter Jet Built with the Cooperation of US Firms," *Asian Wall Street Journal Weekly* (New York), 19 December 1988, p. 16.

16. AFP (Hong Kong), 14 November 1988, in FBIS-CHI 14 November 1988, p. 60.

17. See note 3.

18. See "Moscow's New Tack," *Far Eastern Economic Review* (Hong Kong), 14 August 1986, p. 7.

19. See *New York Times,* 17 September 1988, p. 1.

20. See Steven M. Goldstein, "Diplomacy amid Protest: The Sino-Soviet Summit," *Problems of Communism,* September–October 1989, pp. 49–71, for an analysis of this meeting.

21. Gerald O'Rourke, *Proceedings,* April 1989, pp. 79–83.

22. XH 9 March 1989, in FBIS-CHI 10 March 1989, p. 19.

23. See, for example, Jonathan Moore, "President Lee's State Visit to Singapore Wins Kudos: Flexible Diplomacy," *Far Eastern Economic Review,* 23 March 1989, pp. 28–29; Lin Ching-wen, "Lee Man of Hour by Any Name," *Free China Journal* (Taipei), 13 March 1989, p.1.

24. *China Daily: Business Weekly Supplement* (Beijing), 23 March 1989, p.1; see also Steven Erlanger, "China and Vietnam Try Some Trade Diplomacy," *New York Times,* 15 April 1989, p.4.

25. *Yonhap* (Seoul), 6 April 1989, in FBIS-EAS 6 April 1989, p. 35.

26. Choe Won-sok, "Businesses Welcome Direct Trade with USSR," *Korea Times* (Seoul), 6 April 1989, p. 8.

27. Norman D. Palmer, "United States Policy in East Asia," *Current History,* April 1989, p. 181.

TWO

China and Asian Security

Donald S. Zagoria

hina's foreign policy in the 1990s will be determined by
three major factors: the radically changing global strategic
environment, particularly the winding down of the cold
war, which has been at the center of international politics since the
end of World War II; China's continuing national priorities: modernization, security, and reunification with Hong Kong and Taiwan; and, finally, China's domestic politics since the Tiananmen
massacre, a politics characterized by the triumph of the hardliners
over the reformers, a ruthless repression of dissidents, and a new
drive towards reviving central control over the economy.

In this essay, I want to consider each of these factors. The first
section discusses the changing international environment and its
likely impact on Chinese foreign policy. A second section discusses
the high-priority national interests that have provided considerable continuity to Chinese foreign policy during the past decade:
modernization, security, and reunification with Hong Kong and
Taiwan. A third and concluding section discusses the post-Tianan-
men domestic policies of China's new conservative leadership, the
high costs that China has paid for these policies, and some potential scenarios for the future.

The Changing Global Strategic Environment

Some forty years after the beginning of the cold war, the world
is on the verge of a new era, one that promises more stable,
peaceful, and even cooperative relations between the two superpowers that have dominated the postwar era. Within reach is a
world in which both the Soviet Union and the United States make
radical cuts in conventional and nuclear arms; the division of

Europe comes to an end as Eastern Europe, loosed from the grip of Soviet power, moves towards a market economy and a multiparty system; an increasingly prosperous East Asia becomes more stable and peaceful as the major powers in the region approach detente with each other; regional conflicts in the third world are resolved or attenuated; and greater attention is paid to collective global concerns such as the environment, pollution, poverty, drugs, and so on.

Of course, such a momentous change in global international relations is by no means assured. Its outcome depends on many factors, not the least of which is the positive outcome of Mikhail Gorbachev's *perestroika* in the Soviet Union. Still, for the first time in many decades, it is now possible to see the outlines of a more stable, rational, and prosperous world.[1]

One of the principal factors contributing to the prospects for a more peaceful world is ironically the economic and sociopolitical crisis now confronting the Soviet Union and the domestic and foreign policy changes that Gorbachev has been forced to make as a result of this crisis. In order to cope with its domestic problems, Moscow has been forced to release its hold on its East European empire, to withdraw from Afghanistan, to cut its troop strength and its defense budget, to make a number of concessions to the West on arms control, and to come to terms with all of its former adversaries.

There are at least three important factors that drive this revolution in Soviet foreign policy. The first is tactical. Gorbachev inherited a foreign policy from Leonid Brezhnev that was extremely counterproductive. Brezhnev's rigid, dogmatic, and overmilitarized foreign policy, culminating in the deployment of the SS-20 intermediate-range nuclear missiles and the Soviet invasion of Afghanistan, was leading by the early 1980s to a united front of all the major power centers against the Soviet Union. A Chinese-Japanese-European-American alliance loomed on the horizon. Any post-Brezhnev Soviet leader would have wanted to take steps to break up the powerful combination of adversaries that Brezhnev's expansionist foreign policy had brought about.[2]

A second and more fundamental reason for the revolution in Soviet foreign policy has to do with the fact that Gorbachev's highest priority is to reform and to modernize the stagnant Soviet economy. Unless he succeeds in this endeavor, the Soviet Union will become a second-rate power, and its government will face a continuing crisis of legitimacy at home. But for reform, the Soviet leader needs a peaceful international environment, arms control agreements with the United States that can justify a drastic reduction in the Soviet military budget, and an integration of the Soviet economy into the global economy so that the Soviets can gain access to Western credits, trade, expertise, and technology.

A third reason for the revolution in Soviet foreign policy is that Gorbachev realizes that in the modern world there are substantial limits to the use of military power and that the very nature of power is changing. In the age of the information revolution, progress in science and technology is as important a key to power as were vast arms buildups and territorial aggrandizement in earlier eras.[3]

In sum, there are very powerful forces behind the new moderation in Soviet foreign policy, and even if Gorbachev were to be replaced by a more conservative Soviet leader, it is likely that the Soviet Union would still need a more benign and less threatening international environment in order to concentrate on its domestic agenda.

In addition to the revolution in Soviet foreign policy, another critical factor shaping the new global strategic environment is the collapse of communist regimes in Poland, Hungary, East Germany, Romania, and Czechoslovakia. There is now a good chance that most of these Eastern European communist states will move towards market economies and multiparty systems in the near future and that they will enter into a more cooperative relationship with the European Economic Community.

As a result of the changes in Soviet foreign policy and the breakdown of communism in Eastern Europe, there are now prospects for ending the division of Europe and for unifying Germany. Pressures are building for a very large reduction of Soviet forces in Eastern Europe and of American forces in Western Europe. The Soviets have already begun to cut back on the numbers of tanks and mobile forces stationed in Eastern Europe and thus to relieve NATO of its long-standing concern about the threat of surprise attack.

Attitudes are also changing on both sides of the East-West divide. In the West, where the Soviets were once seen as unalterably belligerent, militaristic, and expansionist, they are now viewed as mired in a protracted crisis and considerably less threatening. On the Soviet side, there is a growing awareness that we are all entering a multipolar world in which the United States and the Soviet Union may increasingly come to share some common concerns such as nuclear proliferation.

Moreover, in both East and West, fiscal constraints are forcing a reduction of defense budgets. Gorbachev has announced a cut in defense spending of some 15 percent in the next few years, and the Bush administration has outlined a massive cut of some $180 billion in defense spending over the next several years. Some independent analysts are calling for a cut of 50 percent in the United States defense budget by the year 2000.

In sum, the demise of communism in Eastern Europe, the revolution in Soviet foreign and domestic policy, the fiscal and economic constraints in East and West, as well as changing Western and Soviet perspectives, are all leading to an end of the division of Europe and to an attenuation of the cold war.

Although there are important differences between the strategic situation in Europe and that in Asia, the ending of the cold war in Europe is bound to have a calming effect on great-power and regional rivalries in Asia.

A Sino-Soviet detente is already under way, and it is likely to last for some time because both the Soviet Union and China need a long breathing space while they concentrate on rejuvenating their economies. Gorbachev met Deng Xiaoping in May 1989, the first such meeting between top Soviet and Chinese leaders in thirty years; and a normalization of Sino-Soviet relations, including mutual force reductions, a resolution of the border conflict, and increased trade and contacts are all likely to follow. The Soviet Union has already pledged to withdraw 200,000 troops from Asia within the next several years, and most of these will come from Mongolia and the Sino-Soviet border. China has made substantial cuts in its troop strength, at least on paper, and is reportedly planning further cuts.

As a result of their detente with each other and their continuing preoccupation with modernization, both the Soviet Union and China will have a growing stake in peace and stability in Asia. Both are now seeking to calm troubled waters in the region by urging North Korea to compromise with South Korea and by seeking a diplomatic solution to the Cambodian conflict.[4]

Soviet relations with Japan are also improving. Gorbachev is scheduled to visit Japan in 1991, and a number of Soviet academics and diplomats are suggesting possible compromise solutions to the territorial issue that has blocked their relations for many decades. The Japanese, for their part, have reported progress in drawing up a peace treaty with the Soviets and have indicated, following the Malta summit between Bush and Gorbachev, that there are now "real possibilities" for improving Japanese-Soviet relations.[5]

Chinese-Indian relations are also showing some signs of progress since Rajiv Gandhi's visit to China in October 1988, the first such trip by an Indian prime minister since Nehru visited in 1954.

Although Chinese–United States relations went into a tailspin after the Tiananmen incident, both Washington and Beijing have been reluctant to allow the deterioration to get out of hand. President Bush sent high-level missions led by National Security Adviser Brent Scowcroft to China in July and December 1989, and the Chinese lifted some restrictions on the Voice of America and agreed, at least in principle, not to sell missiles to Middle East countries.

Another encouraging trend in East Asia, and one that is bound to influence future Chinese policy, is the continuing economic dynamism of the region and the increasing tendency towards regional economic cooperation. Intra-Asian trade is growing rapidly. Trade and foreign ministers from twelve Asia-Pacific countries, including the United States, met in Canberra in November 1989 for the first regional conference of Pacific ministers. The Asia-Pacific Economic Cooperation (APEC) Conference was attended by representatives from Australia, Canada, Japan, Korea, New Zealand, the United States, and the six ASEAN countries. The Canberra conference represented a major step forward in the area of Pacific cooperation. Discussions centered on four principal topics: world and regional economic developments, the role of the Pacific in trade liberalization, specific opportunities for regional cooperation, and the future development of the organization.[6]

In the 1990s and into the twenty-first century, East Asia is likely to emerge as one of the leading economic engines of global transformation. East Asia already produces close to 20 percent of world GNP, not far behind North America, which now produces 27 percent. By the end of this century, East Asia will probably contribute as much as North America to world GNP.[7]

This economic dynamism has been accompanied by a restoration of democracy in the Philippines and a new push towards democracy in South Korea, Taiwan, and Pakistan. To be sure, democracy is still a fragile institution in most countries in the region, but there is a general tendency towards greater pluralism. Economic development often generates pressures for political pluralism.

Another factor that has contributed to peace and stability in East Asia—and the major subject of this paper—is the dramatic change in China's orientation. China has abandoned the revolutionary domestic and foreign policy it pursued under Mao Zedong and has begun to focus on modernization as its primary objective. As a result, China has joined the Asia-Pacific trading system and the global economy. So long as China is absorbed with domestic modernization, it will have a continuing stake in a calm international atmosphere and in obtaining Western credits, expertise, and technology.

There is one final factor in the newly emerging global stategic situation to which I want to call attention. Despite much talk about the "decline" of the United States, every one of America's allies in Europe and Asia wants to see a continuing American military presence on the Eurasian continent.

America's NATO allies want the United States to remain in Europe until there is much greater progress towards European integration and a resolution of the newly revived German reunification issue. In Asia, China and Japan want the United States to continue balancing Soviet power; South Korea wants a strong United States presence to deter and to balance North Korea; and all of the ASEAN countries want the United States to remain in Asia as a counterweight to a modernizing China and a potentially resurgent Japan.

In the new multipolar world that is emerging, even the Soviet Union is coming to have an interest in the maintenance of the American forward deployment system. In Europe, Moscow is no longer calling for the abolition of NATO and the Warsaw Pact because it now recognizes that the continuation of these two alliances will be essential to guide Europe towards a new security system that will anchor a reunified Germany. And in Asia, while Moscow is ambivalent about the United States–Japanese alliance, it is now coming to realize that a Japan cut loose from its American connection may be more threatening to Soviet interests than a Japan that is safely tied to the United States.

The new global strategic environment is bound to influence China's foreign relations in many ways. First of all, provided the international environment remains relatively benign, there will be no incentive for China to make any substantial increases in its military spending, which has been on a steady decline in recent years. Second, the substantial improvement in Soviet–United States relations will limit China's ability to play off the superpowers against each other. For both Moscow and Washington, relations with each other will now be much more important than their respective relations with China. The primary objective of each of the two superpowers will be to construct a new European security system in the wake of the demise of the old one, to reach nuclear and conventional arms control agreements that will permit a reduction of defense spending, and to calm regional tensions that could flare up and threaten their new detente. Thus, for the United States, at a time when the Soviet threat seems to be diminishing, there will be much less incentive to play the "China card" against the Soviets. Although China will remain an important regional power for the United States, China will lose its centrality as a factor in shaping American foreign policy. For Moscow, at a time when good relations with the West are essential for ensuring the success of *perestroika*, there will be considerably less incentive to try to use better relations with China against the United States. And for Moscow, too, China will become less central in its foreign policy considerations as Europe and the United States assume a much larger role in the future of *perestroika*.

Moreover, at a time when economics, not military power, is becoming such a dominant force in international relations, China will continue to be constrained by its weak economic and technological base and by its need to join the international trading system. Even China's new hardline leaders recognize that, if they want to modernize, there is no alternative to importing Western technology.

China's Foreign Policy Priorities

In addition to the international strategic environment, another major variable influencing China's foreign policy will be Beijing's national priorities. During the past decade or more, these have remained fairly consistent, and as a result there has been a remarkable continuity in Chinese foreign policy. China has desired

- A growing integration with the world economy so as to support its ambitious plans for economic development;
- A peaceful international environment, particularly in Asia, so that it can minimize its military expenditures and devote the greatest possible attention to domestic affairs, while forging beneficial economic relations with the widest range of countries;
- A stable relationship with both superpowers, avoiding confrontation or alignment with either, while pressing both Moscow and Washington to remove what China has defined as the major "obstacles" to an improvement of relations;
- Progress towards the reunification of Hong Kong and Taiwan with the People's Republic of China.[8]

Although Chinese foreign policy has undergone some changes since the Tiananmen tragedy, and despite the cooling of relations between China and the West as a result of the crackdown on the pro-democracy movement, there is nothing to suggest that Beijing has lost interest in any of the four goals outlined above, or that it is planning to return to the revolutionary foreign policy it conducted during the Mao era.

So far as integration into the world economy is concerned, Chinese foreign trade over the past decade has grown remarkably, from around $15 billion in the mid-1970s to more than $80 billion in 1988.[9] As a result, China has become the fifteenth largest trading country in the world.

China's economic relations with Japan have become particularly important to the success of China's modernization. Japan has in just a decade become China's second largest trading partner, a principal provider of financial and technical assistance, and a major market for China's energy and labor-intensive exports. After the signing of the Sino-Japanese Long Term Trade Agreement in 1978, the volume of trade between the two countries increased from $5.1 billion in 1978 to some $20 billion in 1988. The new austerity program introduced by the Chinese leaders to cool off the economy may lead to a decline in trade with Japan as well as in foreign trade more generally, but economic relations with Japan will continue to occupy a high priority in Beijing.

Japan's role in China's economic development has become critical in at least four respects. First, Japan has provided substantial financial assistance to China. Japanese credits, now amounting to more than $4 billion in Ex-Im Bank loans and a similar amount in official development loans, have played a key role in financing the construction of thirteen industrial projects and in the development of China's coal and crude oil resources. Since the events in Tiananmen, Japan has put a hold on loans to China, but there are some signs that Tokyo may be reconsidering this policy because of its desire not to isolate China.

Second, Japan has been the primary source of China's imports of industrial plants. Since 1978, approximately 48 percent of China's imports of whole plants have come from Japan, and this has been one of China's principal means of obtaining high technology.

Third, since 1981, Japan has played an important role in renovating existing factories in China and in training Chinese technical personnel. Finally, the Japanese have set up some 183 joint ventures in China.[10]

China's economic relations with the United States, though they have been running into many difficulties since Tiananmen, are also crucial for Beijing. By the spring of 1989, prior to the crackdown, annual United States–China trade had topped $8 billion, and the United States had become the second largest investor in China, after Hong Kong. Several hundred American companies— among them Heinz, Chevron, Occidental Petroleum, Babcock and Wilcox, American Motors, Hewlett-Packard, Otis Elevator, and Squibb Pharmaceuticals—had committed a total of $3.5 billion to ventures run jointly with the Chinese.[11]

In recent years, the United States has also been assuming increasing importance as one of China's most important export markets. The United States is China's largest market for clothing and the third largest purchaser of Chinese yarns and fabrics. Light industrial goods are also becoming an ever larger share of China's sales to the United States. Exports of sporting goods, toys, travel goods, handbags, footwear, and tape recorders have all been increasing rapidly.

Moreover, the United States is an important actor in international economic organizations such as GATT and the World Bank, organizations with which the Chinese are anxious to associate. The Chinese want to join GATT and to obtain a resumption of loans from the World Bank and other international agencies. These loans were suspended after Tiananmen.

China's economic relations with other countries of the Pacific Rim aside from Japan and the United States are also increasingly critical to the success of China's modernization. Prior to the Tiananmen crackdown, trade between the People's Republic and South Korea had reached the level of around $2 billion per year despite the absence of formal diplomatic relations between the two countries. The South Koreans expect a substantial increase in trade with China, and they have begun to develop new ports on their west coast directly opposite China.

China's trade with Taiwan is also expanding. Two-way trade, still largely indirect, now amounts to some $2 billion per year. China's trade with the ASEAN community is also expanding. China-ASEAN trade moved from a mere $300 million in 1972 to some $4.2 billion in 1985. This China-ASEAN trade is likely to continue to grow even more rapidly in the years ahead, particularly now that China and Indonesia are about to resume diplomatic relations.

While China increases its trade with the market economies of the West, it is also increasing its trade with the Soviet Union and with Eastern Europe. The Soviet Union has become China's fifth largest trading partner in recent years, behind Japan, Hong Kong, the United States, and West Germany. In 1988, Sino-Soviet trade topped $3 billion. Border trade has been expanding with particular rapidity. China and the Soviet Union are also talking about a variety of joint ventures in Siberia.

In sum, over the past decade, as a result of its open door policy and its market reforms, China's foreign trade has grown greatly and China has derived considerable benefits from its increasing participation in the global economy. Although there has been retrenchment and even backtracking on the reform process in the past year as China's new hardline leadership has tried to re-centralize its economy, cut inflation, and slow down overheated economic growth, China has not cut back substantially its participation in the global economy. Before the Tiananmen events, China's Seventh Five-Year Plan (1986–90) called for a 40 to 50 percent increase in foreign trade and for a substantial increase in foreign investment and borrowing during the same period.

To be sure, in the aftermath of Tiananmen, foreign investors are going to be wary of embarking on new ventures in China, particularly if some of the disturbing tendencies of the past few months continue.[12]

But despite these alarming trends, it is difficult to imagine China's reverting to the isolationist and anti-Western policy of the 1960s. China's open door policy has contributed to a doubling of China's GNP, to a substantial improvement in the living standards of the Chinese people, and to the acquisition of a good deal of foreign technology that is indispensable if China is to achieve its stated goal of becoming a modern industrial power in the next century. Moreover, the Chinese themselves have attributed much of their success in the past decade to their open door policy. Also, even the most conservative of the Chinese leaders continue to stress their intentions to continue this policy.

Thus, although there may be some slowdown in the growth of foreign trade as the retrenchment campaign continues, and although foreign investors may be increasingly cautious about committing new funds to China, it seems likely that China's integration into the Pacific regional economic community will continue, albeit at a slower pace than before the Tiananmen crackdown.

If pursuing modernization through integration into the global economy has been one high priority for China during the past decade, pursuing Chinese security interests by balancing itself between the two superpowers has been another. Indeed, Chinese analysts carefully monitor the competition between the superpowers, the changing balance of power between them, and the implications this balance has for China.[13]

To be sure, as I have already stressed, the substantial improvement in relations between Moscow and Washington will limit Beijing's leverage over both as we enter the 1990s. Still, the most effective game for a relatively weak China is the one it has been playing since 1982—balancing between the superpowers while aligning with neither and seeking to manipulate the Soviet–United States rivalry for Chinese interests.

So long as China shares a 4,600-mile-long border with the Soviet Union, so long as Soviet military power far exceeds that of China, and so long as the Soviet Union maintains military ties to a variety of Asian countries such as North Korea, Mongolia, Vietnam, and India, the United States and China will continue to have a common interest in offering a counterpoise to Soviet power in Asia. In the past, this common interest has been reflected in parallel American and Chinese policies towards Afghanistan, Cambodia, Pakistan, and Thailand. It is manifested also in China's continuing acquiescence to the presence of American military bases in the Asia-Pacific region, and especially to America's alliance with Japan, the cornerstone of American power in the Pacific.

The shared American and Chinese interest in balancing Soviet power is also reflected in China's continued willingness to allow the United States to monitor Soviet missile flights from Chinese territory.

Chinese analysts consider that Moscow's greatest strategic weakness is its need to fight on two fronts if it comes to war. To exploit this weakness, Chinese security interests dictate the maintenance of sufficiently stable ties with the West that Moscow will be deterred from increasing pressure on either its eastern or its western front out of fear that it will have to face adversaries on both. This is a game that might be called "two front deterrence." It was this logic that brought Mao Zedong and Richard Nixon together in 1972 and that has sustained the Sino-American relationship—along with economic considerations—ever since. This common interest is likely to remain.

There are, however, a variety of factors that will contribute to a limited improvement of Sino-Soviet relations in the coming years. First, both China and the Soviet Union will continue to be preoccupied with internal economic reforms that could take years, and perhaps decades, to implement. Each country wants a stable international environment to facilitate domestic reform.

Second, several of the major irritants in the Sino-Soviet relationship are either gone or are being removed. The Soviets have withdrawn from Afghanistan, and the Vietnamese, under Soviet pressure, have withdrawn most of their forces from Cambodia. Moscow has also begun to substantially reduce its forces on the Chinese border.

It is likely that in the next few years there will be a border agreement, more summit meetings, the restoration of party-to-party relations, and some increase in trade and cultural exchanges.

But Sino-Soviet relations will continue to be constrained by a lack of mutual trust, by geopolitical competition in Asia, and by the imbalance of military power. Moreover, another constraint on Sino-Soviet relations after the Tiananmen crackdown will be divergent attitudes towards reform. Unlike the Soviets, who have made positive statements about recent democratic and market reforms in Eastern Europe, the Chinese have condemned these developments as running counter to true socialism.

After modernization and security, reunification of Taiwan and Hong Kong with the mainland is the third major priority of Chinese foreign policy.

According to the 1984 Sino-British Joint Declaration, Hong Kong will be returned to Chinese rule in 1997. Any Chinese government would presumably wish to retain as much as possible of the present system in Hong Kong in order that its dynamic economy could contribute to China's own economic success. Since the Tiananmen incident, the Chinese have made a number of statements designed to reassure Hong Kong that Beijing plans to live up to its promise to keep a capitalist system in Hong Kong for the next fifty years. But since the June crackdown, China has also been intensifying its attacks on the British government's handling of the Hong Kong issue and ignoring the detrimental effect its new hard line is likely to have on confidence and stability in the colony. The Chinese have been particularly critical of Britain's plan to offer limited nationality to some Hong Kong British subjects, and to introduce political changes in the colony designed to give it more representative government.

The Chinese hardliners, intent on suppressing any remnants of the democracy movement inside China, are concerned over the prospect that Hong Kong could be used as a base to subvert the central government in Beijing and to undermine their efforts to restore some degree of Marxist orthodoxy on the mainland.

If the Beijing government continues its harsh policies of restoring central control of the economy and repressing all signs of political pluralism, it is likely to breed a growing lack of confidence on the part of Hong Kong. Already large numbers of Hong Kong professionals are seeking permanent residency abroad.

Relations between the People's Republic and Taiwan are also likely to be adversely affected by the June 4 bloodletting. In recent years, there had been modest improvement in relations between

Beijing and Taipei, leading some observers to believe that a rapprochement between the traditional foes was on the way. Two-way trade is growing; so is Taiwan investment in the mainland. A large number of Taiwan residents was allowed to visit the mainland and a regularized indirect mail service was inaugurated. Taiwan was also ready to expand cultural, athletic, and academic exchanges. But the Beijing crackdown is now likely to make Taiwan warier of closer contacts.

In sum, it seems likely that, with one important exception, the priorities that provided a considerable degree of continuity to Chinese foreign policy during the past decade will remain. China will continue to pursue integration into the global economy, security through balancing between the superpowers, and reunification with Hong Kong and Taiwan. There will not be any dramatic change in Chinese foreign policy as a result of the Tiananmen affair. Still, a new priority is now clearly evident—a perceived need on the part of China's new conservative leaders to roll back some of the reforms, to silence dissent, and to prevent any new disorders by strengthening the grip of the party on the economy and society. As a result of such perceived needs, the period following the crackdown has led to a worsening of relations with the West and to a more strident tone on issues such as Hong Kong. It is to these domestic factors, and how they impinge on Chinese foreign policy, that I now wish to turn.

China's Domestic Politics

The final factor that will influence China's foreign policy in the 1990s is domestic politics. The crushing of the democracy movement, the purge of the reformers, the new wave of repression, the new efforts to recentralize the economy, and the frequent attacks on the West, particularly on the United States, for allegedly instigating the democracy movement and for seeking to export "pluralism" to China are all likely to weigh heavily on China's relations with the outside world. Since Tiananmen, China has lost international goodwill and economic aid; some twenty countries, mainly the Western industrial democracies and Japan, have cut off high-level contacts and frozen loans; there has been a substantial short-term decline in new direct foreign investment in China, and tourism, once a major earner of hard currency, has declined precipitously.

China's relations with the outside world, particularly with the United States, have been placed under considerable stress. Adding to the strain with the United States were specific events such as the fact that Chinese dissident Fang Lizhi and his family sought and obtained temporary refuge in the American embassy in Beijing.

Since June, there has been a wave of arrests and executions, and ideology has come back into the limelight as China has moved backwards towards a kind of doctrinaire socialism. Students have been removed from their studies and sent to the "grassroots" to learn from workers and peasants. New prominence is given to the party in factories and offices. And Chinese leaders have begun to display the xenophobic resentment towards the outside world that characterized much of the Mao era. Foreigners, notably the United States, are blamed for helping to cause the student protests and for isolating China since. The sanctions imposed by Western countries are condemned by China's conservative rulers as interference in China's internal affairs and even as an effort to subvert China's efforts to develop socialism.

Meanwhile, China's political stability looks precarious. Many of those now holding power are in their eighties, and the younger leaders in their sixties and seventies lack the stature to ensure their staying power once their elders are gone.[14]

There are also indications of substantial policy disagreements within the new ruling group, and another succession fight seems inevitable when Deng Xiaoping dies or is incapacitated. All of this means that the central authorities are in a weak position to command the loyalty and obedience of local officials, many of whom resent the new efforts at imposing central control.

The Chinese economy also looks to be in serious difficulty. In its determination to dampen the energized economy, the new leadership is putting on a credit squeeze that threatens to kill rather than simply to retard expansion. Industrial growth has contracted in recent months, and Chinese factories and businesses are shutting down or scaling back. Small private enterprises that helped to breathe life into the economy have been hit hard, and some million of them have reportedly been forced to close. Meanwhile China's debt, though not great by international standards, has reached $40 billion, and China faces a peak in repayments in the early 1990s.

The situation in agriculture is also poor. Rural industry, which was intended to provide a livelihood for surplus farm workers and boost the standard of living, has suffered so tight a credit and raw materials squeeze that thousands of factories have closed. Ominously, Beijing is experimenting with new systems of land allocation that would, if implemented, bite deeply into the post-1979 peasant freedoms. And the grain harvest has fallen off badly after its peak of 497 million tons in 1984.

For its continuing suppression and its backtracking on reform, China is paying a heavy price in foreign relations. In Hong Kong, there has been an enormous erosion of confidence, and the British are under heavy pressure to negotiate stronger safeguards for Hong Kong before reversion to Chinese sovereignty in 1997. China's efforts to woo Taiwan back to the "motherland" are bound to be substantially set back. Japan, though anxious that China not be isolated, has placed its sizable development assistance program on hold; and many of the estimated 8,000 Japanese previously working in China have not yet returned after the crackdown. In South Korea, there is much less optimism about future trade prospects with China. In the Soviet Union, there is anxiety that an unstable China will not be a good neighbor and that a more Stalinist China will be wary of too-close contacts with a "revisionist" Soviet Union.

The costs to Sino-United States ties have been enormous. The suspension of military sales and the decision to encourage the World Bank to hold up loans for China have been severe blows to Beijing. But even more important is the general decline in the confidence with which Americans had viewed China's future. Moreover, China has now once again become a major issue in American politics. Congress wants to take firmer sanctions against Beijing than the Bush administration has so far adopted. Many in Congress and in the press are extremely critical of the Scowcroft missions to China as an unwarranted concession to such a repressive government.

There are three possible scenarios for the future evolution of Chinese politics. First, the conservatives may succeed in consolidating their power and in rolling back some of the reforms. But if this happens, China is likely to face a lengthy period of Brezhnev-type stagnation that almost inevitably will trigger new political and social instability at some later date. Second, the reformers may yet make a comeback, particularly if the conservatives fail to get the economy moving again. Finally, China may be increasingly confronted with pressures from below similar to those that helped remove communist regimes in Eastern Europe.

In the short term—the next two or three years—the most likely scenario is that China's conservative leaders will consolidate their power, continue with their efforts to recentralize the economy, and go on with their ideological campaign against "bourgeois liberalization." Under these circumstances, there is little likelihood that there will be any substantial improvement in China's relations with the Western industrial democracies and Japan.

But over the longer run, the underlying pressures for reform in China will remain and even grow stronger. Moreover, China cannot remain immune from the pressures that are forcing radical changes in communist governments from Budapest to Berlin to Moscow. At some point, we are likely to witness a new push towards reform in China.

NOTES

1. See Robert D. Hormats, "The Economic Consequences of the Peace—1989," *Survival*, November–December 1989, published by Brasssey's for the International Institute of Strategic Studies, London.
2. Donald S. Zagoria, "Into the Breach: New Soviet Alliances in the Third World," *Foreign Affairs*, Spring 1979.
3. See Richard Rosecrance, *The Rise of the Trading State* (New York: Basic Books, 1986).
4. See Donald S. Zagoria, "Soviet Policy in East Asia: A New Beginning?" *Foreign Affairs, America and the World*, 1988–89.
5. See the statement by Japanese Foreign Minister Taro Nakayama in FBIS-EAS 11 December 1989, p. 2, and the statement by Prime Minister Toshiki Kaifu in FBIS-EAS 8 December 1989, p. 1.
6. See the PBEC Bulletin, Canberra Ministerial Conference, 6–7 November 1989, The Pacific Basin Economic Council, San Francisco.
7. See the essays in James Morley, ed., *The Pacific Basin*, Proceedings of the Academy of Political Science, vol. 36, no. 1, New York, 1986.
8. Harry Harding, "Recent Trends in Chinese Foreign Policy," May 1988, unpublished manuscript.
9. See Nicholas Lardy, *China's Entry into the World Economy*, Asian Agenda Report 11 (New York: Asia Society, 1987).
10. Hong N. Kim, "Sino-Japanese Relations," *Current History*, April 1988.
11. See "China Has Been Giving U.S. Businesses the Cold Shoulder," *Washington Post*, National Weekly Edition, 13–19 November 1989, p. 22.
12. In October 1989, Chinese authorities abruptly ended the allocation of foreign exchange for one Chinese-American joint venture and told another to stop manufacturing for the domestic market. Still another was told that its annual profit would be limited in the future to 3 percent of gross earnings. Such actions sent a shock wave through the foreign investment community. See ibid.
13. See Banning Garrett and Bonnie Glaser, "Chinese Estimates

of the U.S.–Soviet Balance of Power," occasional paper no. 33 (Washington, D.C.: Wilson Center, 1988).

14. See David M. Lampton, *China and U.S.-China Relations in the Wake of Tiananmen*, National Committee on U.S.-China Relations, China Policy Series, September 1989, a report of a conference held 7–9 July 1989; see also the *London Financial Times* supplement on China, 12 December 1989.

THREE

Contemporary United States– Japan Security Relations: Old Myths, New Realities

Donald C. Hellmann

The Japanese-American alliance is one of the most successful and enduring achievements of postwar American foreign policy, but it is doubtful that it can continue in its current form in the decade ahead. The reasons mandating change are not found in the day-to-day military cooperation between the two countries, which has proceeded smoothly and well. Rather, policy change is required by the enormous and continuing structural shifts in the global and East Asian political economies. Accordingly, to understand security relations between Japan and the United States and the unfolding policy challenges, it is necessary to look not only at the military capabilities of each nation and the institutions designed to foster defense cooperation, but to the new international economic and political realities within which the alliance must operate.

Japan has become a global economic superpower, of proportions that are rarely fully understood. Its gross national product has been larger than that of the Soviet Union for several years, and its per capita GNP is more than 20 percent larger than that of the United States. Moreover, compensating for vastly inferior natural resources, Japan has used highly effective government-private sector institutional arrangements to maximize economic growth, producing a lead in many high-technology areas that provides momentum to sustain rates of growth greater than those of the United States and the Soviet Union. It is the Japanese model, not American capitalism or Soviet socialism, that is the most admired international prototype for economic growth. Japan has become the world's foremost financial power: the largest international creditor, with growing overseas investments; an aid program that in 1989 will be the world's largest; a strong banking system and an extremely high personal savings rate (four to five times the rate in the United States). At the same time, the United States has stumbled, developing huge deficits in its national budget and in trade, the largest (and growing) international debt of any nation in modern times, a fragmented and shaky banking system, the lowest rate of domestic savings of any industrialized nation, and a serious problem of failing international competitiveness in many industrial sectors. The one-sided nature of the relative economic performance of Japan and the United States over the past decade and a half has also caused the American and Japanese economies to become ever more deeply intertwined.[1] The sheer degree and extent of interdependence has made the bilateral relationship the largest single factor in maintaining global economic stability; at the same time, it has exacerbated the problems of managing bilateral macroeconomic relations and raised doubts about the long-term viability of the one-sided, American-dominated security relationship.

Despite its enormous accretion of economic power, Japan has continued to play a passive and modest role in both the security and economic dimensions of the international system. Economically, Japan has worked almost exclusively to maximize its own national self-interest. In international institutions such as the GATT (General Agreement on Tariffs and Trade), designed to promote and strengthen the free trading system, Japan has been a major target of criticism rather than a leader. Although for several years the world's largest creditor nation, Japan has left the initiative for solving the third world debt to the United States—the world's largest debtor. Fourteen of the fifteen largest banks in the world are now Japanese, but through the mid-1980s the partial opening of the financial markets of Japan occurred only because of intense international pressure. Despite a significant increase in the magnitude of its foreign aid (that remains de facto over 60 percent tied to the Japanese economy), Japan has shown little inclination or capacity for international economic leadership beyond maximizing its own national wellbeing.

In the area of security, the role of Japan has been even more modest. As elaborated in the National Defense Program Outline adopted in 1976, the strategic aim is narrowly defined to involve territorial defense of the home islands.[2] Even with this restricted goal, the military role of Japan is seen as simply a supplement to the conventional military capacities of the United States. Japan is sheltered by the American nuclear umbrella, but unlike NATO nations, under the so-called three non-nuclear principles Japan will not permit transshipment or deployment of the American nuclear weapons used to provide an umbrella in the Western Pacific. Because of the American alliance and these highly restricted strategic aims, Japan spends (and has spent) only one percent of its GNP on defense—an amount that in proportional terms is less than 25 percent of the average annual outlay of any member of NATO.[3] Japan not only eschews a regional security role, but alone among the major industrial powers has not participated in the United Nations peacekeeping activities or made a military contribution to the multilateral efforts to ensure access to Middle East oil.

This profound and extreme incongruity between the enormous and growing economic power of Japan and its highly restricted role in the management of both the economic and political-security elements of global affairs can be extended into the future under only one condition: the United States must remain willing and able to sustain the international greenhouse that has insulated Japan from the political, psychological, and economic costs and responsibilities associated with an orthodox national role in global defense arrangements. This greenhouse has also allowed Japan to avoid a role of economic leadership broadly congruent with its international economic capacities. Both the capability of the United States to maintain the greenhouse and the desirability of doing so are now open to question. Clearly, it defies both history and common sense for the world's largest debtor nation effectively to underwrite the security of the world's largest creditor. In the future, security issues must be linked with economic issues in Japanese-American relations, especially in light of fundamental changes that have occurred in the strategic landscape of East Asia.

Most analyses of United States–Japan defense relations do not go much beyond an updated inventory of the inhibitions to Japan's doing anything beyond the modest security role the nation now plays. Focus is primarily on what can be called the five "old myths" about United States–Japan security relations. These are:

(1) Japan's constitutional limits on any change in the current restricted security role, (2) the profound domestic political opposition to expanded Japanese military activities, (3) the deep fear felt in all East Asian nations (and the Soviet Union) at the prospect of a "rearmed Japan," (4) the fact that Japan is already a major military power in terms of the size of its defense budget and the intractability of the bilateral alliance if Japan were to expand its defense spending to a level of any Western European nation, and (5) the need to create "a strategic partnership based on a division of labor in which the Japanese assume major economic burdens to balance American military burdens."[4] These assumptions and hypotheses have guided and still guide the bilateral relationship, but they have long been in desperate need of reevaluation in terms of contemporary international political-economic realities. Without such a reevaluation and appropriate policy adjustments, the bilateral alliance can at best lurch from one crisis to the next. If the policy agenda comes to be set by the drift of international economic and strategic events and the twists and turns of domestic politics in each nation, the alliance that has been central to stabilizing international relations in East Asia will gradually unravel, with grave consequences for the global political economy. The old myths are examined here with an eye to creating a new, viable formula for Japanese-American security relations.

> **Myth #1:** The American-drafted peace constitution precludes the participation of Japan in collective security arrangements or in dispatching troops abroad.

This myth is rooted in article 9 of the 1947 American-drafted Japanese constitution, which denies to Japan in unqualified terms the right to belligerency and to maintaining armed forces of any kind. Reflecting a pacifist mood widely shared in the wake of World War II and the peculiar idealism of the occupation leader, General Douglas MacArthur, this extreme position was immediately challenged by the emergence of the cold war and the outbreak of the Korean War, and the return of Japan in 1952 to the status of an independent nation. Except on the political left, the need to establish some sort of military force was quickly acknowledged in Japan, and in the early 1950s there was even serious debate among the ruling conservatives about constitutional revision. In the end, however, the Self-Defense Forces Act of 1954 was passed and legitimated by appeal to article 51 of the United Nations Charter. Article 51 expressly grants to all sovereign nations "the right to individual *and* collective self defense" (emphasis added). The most critical part of this assertion is not the declaration of the right of individual nations to ensure their survival, but the explicit mention of the universal right to collective self-defense, a point that is underscored by the citation of this article at the beginning of every NATO handbook as the legal basis for the European security arrangement. However, the Japanese have given an astonishingly incomplete interpretation to the article so as to preclude their participation in collective security on *constitutional* grounds.[5]

The reasoning is bizarre but important to understand because it is regularly cited as a cornerstone of Japan's present and future defense policy. Conceding that "as a sovereign state, Japan has the right of collective self-defense," the government nevertheless "is of the view that the exercise of the right of self-defense as permissible under Article 9 of the Constitution is authorized only when the act of self-defense is within the minimum (military force) limit necessary for the defense of this nation. The government, therefore, believes that the exercise of the right of collective self-defense exceeds the minimum limit and is constitutionally not permissible."[6] In this act of verbal legerdemain, a policy choice on the appropriate size of the national military forces is transformed by the Japanese government into a constitutional restraint (a matter of principle) imposed by a document written by the United States occupation forces. Obviously, a policy decision to maintain a military force of sufficient size only to defend Japanese territory is quite different from a legal proscription from doing more.

The ambiguity of this position has on occasion seemingly befuddled the Japanese government itself. For example, in June 1987, Japan flatly turned down a request from the United States to participate in the multilateral naval force sent to the Persian Gulf in the wake of an attack on an American ship, even though (1) the sixty-ship Japanese minesweeping force was at least equal to the capabilities of the United States, (2) Japan received over 60 percent of its oil from the Gulf (the United States obtains only 4 percent), and (3) America's European allies participated, while Japan refused to be involved on *constitutional* grounds.[7] That this clearly was not valid was subsequently acknowledged by the head of the Japanese government. In response to interpellation in the Diet on August 27, 1987, Prime Minister Yasuhiro Nakasone acknowledged that there was no constitutional bar to sending minesweepers, but the self-defense forces would not be sent because "there is a possibility that they will become embroiled (in combat)."[8] In short, the issue of deployment was a policy, not a constitutional, matter. Astonishingly, despite mounting concern in Washington about Japanese defense burden-sharing, no one in the White House, the State Department, the Pentagon, or the Congress chose to challenge the inconsistency in the Japanese position. It is ironic, but the myth of constitutional restraints on deployment of Japanese troops abroad seems to have greater acceptance in Washington than in Tokyo!

> **Myth #2:** The impact on the Japanese public of the catastrophic defeat in World War II, the bombing of Hiroshima, the imposition of the "peace" constitution, and more than four decades on the strategic sidelines in global affairs effectively preclude Japan from becoming an ordinary country in defense. That is, public opinion prevents the Japanese government from spending a proportionate amount on arms and participating in military-security activities, including reciprocal obligations for the United States nuclear umbrella, on a level found in every other nation of consequence in the world.

This myth rests on two basic assumptions: (1) the combination of defeat, imposition of the constitution, and isolation from power politics has resocialized the Japanese people for peace in ways that make them unique among the great powers in modern history, and (2) the manifestation of this popular "mood for peace" in numerous and regular public opinion polls places ineluctable limits on the options for security policy of the Japanese government. Both of these assumptions are prima facie open to serious doubt. To suggest that a country of a hundred and twenty million people has undergone a national personality change ignores the extraordinary international conditions that permitted the isolationist-pacifist policy of the past four decades and flies in the face of both history and common sense. To assert that on the one hand the Japanese government leaders can design and manage a comprehensive international-oriented economic development without equal in the world and that, on the other hand, in matters of national security they are simply *followers* of the national mood, pushes political analysis to the realm of the surreal.

To be sure, there are elements of truth in this second myth. First, it is correct that the shattering impact of total defeat in World War II profoundly affected at least two generations of Japanese. Defeat for Japan was far more devastating than defeat for Germany. The Nazis were an aberration from the mainstream of German culture and Western civilization, a motley collection of the flotsam and jetsam of central European society washed into positions of power by the catastrophes of World War I and the Great Depression. In contrast, defeat for Japan involved discrediting the essence of modern Japanese nationalism, the emperor system, the state religion, and "moral education" that had been centerpieces of the nation's history from the latter part of the nineteenth century. However, it is implausible to argue that the trauma of defeat persists through generations now decades removed from the existential realities of the 1930s and 1940s and after the government has revised school textbooks to tone down Japan's responsibility for the conflict. There is little risk in asserting that generational change has dulled the political impact of the war.

The primary impact of the constitution on Japanese defense policy is political rather than legal (Myth #1). In several ways, it has helped to foster a consensus on a pacifist-tending noninvolvement policy under the umbrella of the American alliance. Since the late 1940s, Japanese left-wing parties and their supporters have embraced the American-drafted peace constitution as a critical weapon in internal politics. For more than two and a half decades, the opposition parties used article 9 (the peace clause) as a moral cudgel to attack both the self-defense forces and the American alliance, in the Diet and in elections. After the debacle in Vietnam (which raised the prospect that the United States might really "go home" from the Western Pacific) and the waning of the cold war in the mid-1980s, the Japanese left initiated a tactical shift, using appeal to article 9 not to foster anti-Americanism but to check the rise of Japanese nationalism with a military component.

The mainstream conservatives have also leaned on the peace constitution to set their foreign policy agenda. In contrast to the left, they have used article 9 to resist American pressures to expand the defense budget. The resulting calculated posture of military weakness and economic strength, which makes imperative the alliance with the United States, is called the Yoshida Doctrine and may be viewed as the grand strategy of postwar Japan.[9] With the fading of the United States as the global superpower and the emergence of Japan as a first-rank world economic giant, there has been a general recrudescence of Japanese nationalism that in turn has provoked modest efforts within the conservative party to amend the constitution. Thus, although the peace element of Japan's foreign policy may have its origins in the constitution, today it is the sheer momentum of success of the Yoshida Doctrine, the absence of a real military threat to Japanese territory, and the willingness of the United States to underwrite Japanese security that impedes any change in policy. The political commitment to peace is conditional.

Beyond the overt use of the peace issue by politicians, public opinion has commonly been given a unique importance in shaping Japan's defense policy—especially the deep support for pacifism and a profound aversion to nuclear weapons (a "nuclear allergy"). There are, moreover, numerous polls extending over the past several decades delineating the history of mass opinion on these matters. However, like polls everywhere, Japanese surveys cannot be taken at face value because of the dubious assumptions regarding political awareness and the connection to political actions on which they rest. For example, to place the undeviatingly high levels of popular support for article 9 and the constitution in proper perspective, it is instructive to look at the state of mass opinion when the escalation of the Vietnam War, China's cultural revolution, and high tension between the United States and the Soviet Union heightened popular concern about defense. A poll taken in February 1967 showed that 60 percent of the people had not read the constitution at all, 34 percent were totally unaware of its contents, and only 47 percent thought it supported peace.[10] It is doubtful that today these numbers would be significantly different. The low level of information of the Japanese public (in line with that found in other nations) suggests extreme caution in ascribing policy implications on the basis of survey data. Japanese opinion polls also tend to take a barefoot and simplistic approach to fundamental defense issues (for example, Do you favor Japan developing nuclear weapons?). Since outside of any international context, it is absurd to expect that more than a few people will favor nuclear weapons or oppose peace in the abstract, it is to be expected that over 80 percent of Japanese consistently fall into this category. Accordingly, it is not too surprising that the "nuclear allergy" manifested in the opinion polls is used by the Japanese government to justify the policy of the three non-nuclear principles (barring deployment or even transshipment of American nuclear weapons)—but it may be a revelation to some that the government is implementing a plan that will ultimately result in over 60 percent of the nation's electrical power's being generated by nuclear means with only modest and late-developing public opposition.

Although popular moods in Japan place clear political limits on short-term policies of the government, it is a mistake (Myth #2) to derive long-term conclusions about the international role of Japan from a supposed deep and permanent commitment of the public to a place on the sidelines of power politics—and accordingly to restrict the policy options regarding Japan that are open to Washington. Since the Japanese public resembles other publics in an altered international environment, shifts in the mass mood are highly probable.

> **Myth #3:** The memories of Japanese military imperialism in East Asia are so fresh and frightening and current Japanese economic power is so great that the other countries in the region do not want Japan to play a military role under any circumstances. Therefore, the United States should not pressure Japan to play an expanded part in regional security, and the current American-dominated security arrangements in East Asia must continue.

When presented as a policy option, this myth is distinctive because it finds virtually unanimous support from all of the parties involved—Asians, Americans, Russians, and Japanese. What makes this unanimity astonishing is that the myth itself rests on two radical and extremely questionable assumptions about international reality in East Asia: (1) that Japan is an incorrigible "military-holic" nation, incapable of behaving in a responsible manner as part of a collective regional security arrangement, and (2) that the

United States will continue to operate to maintain security in the Western Pacific within the framework of the past four decades despite the enormous shifts in economic and political power that have occurred. To assert that Japan cannot share the burden of maintaining international order in the same manner as all other world powers is prima facie untenable. If true, it would indeed mandate special costs for Japan's participation in the global political economy. Although it is hypothetically possible for the United States to continue to project military power at a level and in ways comparable to that of the past four decades in the immediate future, to imagine doing so in the long run without regard for the massive changes in the political-economic landscape of East Asia would indeed qualify American policy makers for a position in the front row of the parade that Barbara Tuchman has called mankind's "march of folly."

The persistence of this myth in the face of the questionable assumptions on which it rests is in part rooted in the deep and widespread distrust of Japan throughout East Asia at both the societal and governmental levels. Long-standing Japanese preeminence throughout the region in direct investment, trade, finance, and technology has yielded remarkably little goodwill. There have been sporadic efforts by Japan to implement Asia-centered programs of aid and economic cooperation: the wartime reparations program of the 1950s, a flurry of activity in the mid-1960s (including establishing the Asian Development Bank and taking the lead in the "Aid Indonesia" consortium), the aid program of the late 1970s called the Fukuda Doctrine, and the recent ongoing efforts to double official development assistance. However, these efforts have not enhanced the status of Japan as much as the sheer magnitude of activity suggests, for two basic reasons: (1) the overwhelming amount of Japanese aid has served their commercial purposes far more than the social, economic, and political needs of the recipient nations, and (2) Tokyo has explicitly eschewed any supplementary political role to justify and/or enhance the aid program.

In addition to aid, the Japanese have been provided with opportunities to attenuate the highly negative legacy still lingering from their behavior in World War II and a postwar image that for many resembles that of a political, very rich international counterpart of Ebenezer Scrooge (that is, we need them, but we do not like them). In particular, Japan could have taken the lead in addressing the tragic circumstances of Vietnamese political refugees and "boat people" (over 70 percent of which were picked up by Japanese ships), and as the world's largest creditor, it has had the opportunity to underwrite at least some of the enormous amounts of funds needed for the land reform program of Philippine President Corazon Aquino. In both cases, Japan ignored relatively low-risk opportunities to demonstrate leadership divorced from a narrow concern for self-interest. After severe international criticism, the Japanese agreed to fund for several years the UN refugee program in Southeast Asia, and over the past decade, five to ten thousand Vietnamese have been absorbed into Japan. Tokyo has made some gestures toward the Philippines, but again with a caution so extreme that any goodwill from their actions is effectively lost.

Much of the Asian ill will toward Japan is held over from the extreme actions of the Japanese military during World War II and from the harsh colonial policies in the first half of this century in Korea, China, and Taiwan. Remarkably little has been done to soften this image, which persists three generations after the events occurred. Indeed, the Japanese government has contributed to the persistence of resentment by rewriting its school textbooks to tone down the grosser wartime and colonial atrocities (such as the rape of Nanking)—to the outrage of all of East Asia. The government has repeatedly said that Japan will never again become a military power in order to guarantee that it will never again do what it did before—a double-edged statement proclaiming a commitment to a nonmilitary policy because any move toward a broader security role could well lead to irresponsible "military-holic" behavior. These policies have done little to dim the image of Japan as a narrowly nationalistic, even predatory, country. Not surprisingly, despite increasing economic dominance in the region, no nation has ever suggested that it wanted Japan alone to take on truly significant political or economic leadership in East Asia. This gap between Japan's growing international power and its negative image as an international leader will be an increasing handicap in the years ahead.

Nevertheless, none of the East Asian states has completely closed the door on an expanded Japanese military role so long as it is done in close cooperation with the United States or is in the context of a multilateral alliance. Over the years this has been made explicit by several East Asian leaders, but it has never really been part of a policy dialogue because the United States has never set forth a serious proposal for realistically restructuring security arrangements for the Western Pacific in which Japan would be an integral part.[11] Instead, official American policy and virtually all commentary by academics and policy specialists accept the assumptions of Myth #3 and echo the Japanese government's view. For example, a recent article co-authored by two former secretaries of state asserts that "any attempt to deal with the [bilateral trade] deficit by pressing Japan to step up its defense efforts . . . would generate the gravest doubts all over Asia and . . . might deflect Japan from a greater economic contribution to international stability."[12] In essence, this view is an updated version of the "occupation mentality," namely, that the Japanese still cannot be trusted to play a responsible security role beyond territorial defense even though they are our major Asian ally and have the most dynamic economy in the world. By accepting Myth #3, American foreign policy leaders keep the lid closed on any new policies (such as the internationalization of defense of the sealanes from East Asia to the Persian Gulf) at a time when there is a basic international need for innovation and when the issues of defense burden-sharing and bilateral economic relations are increasingly linked in Congress.

> **Myth #4:** Japan should not expand its defense spending because the aggregate size of its defense budget in 1989 will exceed that of all nations except the United States and the Soviet Union.

This myth came into vogue in the late 1980s as the pressure for defense burden-sharing rose in the United States and the drastic revaluation of the Japanese yen dramatically inflated the nominal value of the military budget of Japan. It has been buttressed by some curious distortions in the weapons stockpile of Japan rooted in decades of procurement without any clear link to a realistic and operational strategic purpose. For example, the Japanese now have more destroyers than the American Seventh Fleet—even though Tokyo is behind schedule in implementing the responsibilities for defense of the sealanes up to one thousand miles. Myth #4 represents the logical but absurd extension of the debate over Japanese defense policy, which is rarely focused on specific security problems within an accepted framework involving comparative costs and responsibilities seen in other alliance relationships.

Over the past decade, there has been a steady growth in Japanese expenditures for defense approximating 8 percent per year, but the ratio of defense expenditures to GNP (the commonly accepted standard for comparison of aggregate costs) has still remained steady at one percent. Moreover, beyond the modest and as yet unfulfilled commitment regarding the sealanes, there has been no change in the operational or strategic aims of Tokyo. How then has Japan come to have a defense budget surpassed only by the superpowers? What does this mean? The astronomical relative growth in Japan's defense budget is the direct result of revaluation of the yen, which has in nominal terms increased in value more than a hundred percent since 1985. By doing nothing, the Japanese nominally more than doubled their military budgets over the last few years. It really is little more than an exercise in arithmetic. By giving a purely monetary meaning to burden–sharing, the Pentagon and Congress have allowed this "funny money" approach to distort matters even more in Japan's favor. Former Ambassador Mike Mansfield has endlessly cited the fact that Japan contributed more on a "per soldier" basis to the support of American troops than any other nation in which our soldiers are garrisoned.[13] Given the highest costs for renting land in the world and the high nominal salaries of Japanese workers, factors which are a central part of the formula for burden-sharing, it is hardly surprising that Japan is number one. It also has little significance in strategic terms.

To give meaning to these burden-sharing figures and the size of the defense budget in general is not easy. Obviously, the critical factor for any military expenditure is not the amount spent, but the purpose for which it is expended. To acquire weapons such as destroyers and tanks without relating them concretely to an integrated plan for defense of the Western Pacific and without fully developed organizational structures, is more a weapons procurement program than a serious defense policy. A comparison of the organization and policies related to Japanese defense with the organization and policies for economic development brilliantly devised and implemented by Tokyo over the past four decades, graphically underscores this point. It moves one from the ridiculous to the sublime.

That the Japanese would like to continue the defense policies of the present is understandable in terms of the low marginal costs relative to the enormous benefits. That the United States would wish to do so, citing the nominal magnitude of the Japanese defense budget as justification, is to move to an Alice-in-Wonderland world where Myth #4 makes sense.

> **Myth #5:** The appropriate revision of the current United States–Japan alliance is a new strategic partnership based on a division of international labor in which the Japanese assume major economic burdens to balance American military burdens.

This myth was effectively endorsed as mutually acceptable official policy in a joint communiqué issued at the end of the summit meeting between President George Bush and Prime Minister Noburo Takeshita in Washington in early 1989. Its policy lineage goes back to the concept of comprehensive security introduced in the late 1970s by Prime Minister Masayoshi Ohira. Stripped of its academic and diplomatic verbiage, comprehensive security is a sophisticated effort to use aid to solidify Japanese access to markets and resources in the third world (especially East Asia) that are essential to the country's continued prosperity and at the same time to avoid any real expansion of Japan's military role.[14] However well this may have served Japanese interests in the past and although it now has the enthusiastic agreement of the United States government, such an international division of labor is inherently not viable and is bad policy from the perspective of Washington.

Under the proposed arrangement, Japan would take on the responsibility of addressing the third world as a kind of calculating Santa Claus, assuming the challenging but benign (and potentially profitable) task of improving the well-being of the world's poor. While Japan plays the good guy, the United States is expected to be the international policeman, an economically costly and politically thankless task. At best, Americans would become the Gurkas of the latter part of the twentieth century and, at worst, inextricably entangled in intractable and bloody conflicts on a continuous basis. A three-year-old in the sandbox would never agree to always playing the bad guy, but, *mirabile dictu,* the United States has agreed to precisely this—without a voice of protest raised from any corner of the American political scene. That we should have agreed to Myth #5 is stunning, but to expect it to be a workable international arrangement in the long run is absurd.

In addition to the fallacious assumption on which Myth #5 is based, there is another equally devastating flaw involved. For economic aid to be a substitute for defense expenditures within the framework of an alliance, there must be effectively an identity of interest underlying the policies of the nations involved. This is simply not the case. The purpose of Japanese aid has been, and continues to be, overwhelmingly commercial in nature, while the purpose of American aid (and in a mirror-image sense, Soviet aid) is overwhelmingly political—for example, to contain communism, to promote civil rights. Obviously, there are corroborative and overlapping interests served by the American and Japanese aid programs, and it is difficult to direct aid to purely political or purely commercial purposes. However, the distinction made here is absolutely essential, and its significance is illustrated by even a casual inspection of the past and contemporary scene in the Pacific. Most conspicuously, and despite the increasing percentage of Official Development Assistance (subsidized and untied) as part of the Japanese aid program, the overwhelming proportion of the aid dispensed is effectively (not legally) tied to purchase of Japanese products. For example, the substantial Japanese aid to China (which was strongly encouraged by the United States until the Tiananmen incident) is a vehicle for assessing and exploiting market opportunities in a rapidly expanding economy; and, because of the procedures mandated for approval of projects, over 90 percent of the money dispensed is spent in Japan.[15] Moreover, despite the virtually unanimous and strong international moral disapprobation and sanctions against Beijing after the June massacre in Tiananmen, Japan was the first of the major powers to restore aid (and trade) and Tokyo pointedly withheld public political judgments about the incident. In the wake of this recent dramatic illustration of the commercial nature of Japanese aid, to equate these expenditures with spending on defense makes sense only in the world of myths in which Japanese-American security relations occur.

Because of a series of scandals, most notably the "Recruit

scandal," an influence-buying scheme that touched the heart of the Japanese political-business establishment in sensational ways, there is a real possibility that a coalition of opposition parties could form the government after the next general election. Whether or not Japan experiences its first change in political regime since the end of the occupation, it is clear that the weakened status of the Liberal Democratic party will impede any major change in foreign policy. A kind of bureaucracy-led policy immobilism will operate at the very moment when bilateral relations with the United States are severely strained over trade issues and as Japan is feeling its way toward a new leadership role in Asia and the world. Short of a massive Socialist-led opposition electoral victory—which would produce discontinuity (and probably crisis) in Japanese-American relations—it seems probable that Japan will move ahead along the same security path demarcated by the five myths discussed here. In a word, the fluidity seen in Japanese domestic politics in 1989, marked by rapid turnover of prime ministers and a sharp rise in the political strength of the Socialist party, does not indicate a revolution in politics in Japan; but by reducing the ability of Tokyo to adjust to changes in the international landscape it will magnify any major security-economic issues in relations with the United States and East Asian nations.

Perhaps the most fundamental issue facing Japan in the years ahead is the question of international leadership. Can Japan become a regional leader? A global leader? Although the issues are complex, it seems singularly unlikely that Japan alone (unlike Britain in the nineteenth century and the United States in the decades after World War II) can set an agenda for even regional, far less global, international affairs. First, leadership would call for greater symmetry between the military and economic components of Japanese foreign policy. This, in turn, would require a *perestroika* in domestic political institutions and a redefinition of Japanese nationalism beyond economic growth. Revolutions of this magnitude are not internally generated in countries at the apex of their success. Secondly, the breakdown of Japanese-American relations and/or a revolutionary upheaval in the international environment in East Asia would *inter alia* have a devastating impact on Japanese prosperity, lead to political pressures on Tokyo that would be destabilizing to the current mercantilist-technocratic system, and result in a reshuffling of alignments among the nations of East Asia in which the Americans, the Soviets, and the Chinese (all nuclear powers) would be involved. To assume that the American security greenhouse could collapse without catastrophic effect on the fragile Japanese flower is beyond belief. Finally, being an international leader requires the articulation of a purpose that transcends national power. Throughout its modern history Japan has never displayed in its foreign policy a capacity to transcend its own national interest—in political, cultural, or economic terms. Indeed, the internal campaign to "internationalize" Japanese attitudes is graphic testimony of how far Japan has to go—despite its enormous accretion of economic power.

Since it is improbable that Japan will become a world or even a regional leader in the fullest sense, some sort of restructuring of the Japanese-American alliance is a precondition for stability in East Asia. Above all, there is a need for greater symmetry in the security relations between the two countries. This symmetry involves not only greater balance between Japan's economic capacity and military power, but balance in the respective roles of Japan and the United States in terms of security in the Pacific. The starting point for restructuring is the replacement of the old myths with policies attuned to the present realities.

NOTES

1. See Robert Gilpin, *The Political-Economy of International Relations* (Princeton: Princeton University Press, 1987) for elaboration on the basic transformation that has occurred in the 1980s in the global political economy as well as the extreme economic interdependence in the bilateral relationship—which he partly labels the Nichibei economy.
2. Defense Agency of Japan, *Defense of Japan, 1988* (Tokyo: Japan Times, Inc., 1988), pp. 259–63. Under continuous American pressure since 1981, Japan is now gradually expanding its responsibilities to include securing the sealanes 1,000 miles from its shores.
3. The average annual outlay of NATO members since the inception of the alliance has been 4 to 4.5 percent of GNP. Currently, the average is about 3 percent.
4. George R. Packard, "The Status and Power of Japan," *National Interest* 6 (Winter 1986/87), p. 45.
5. See, for example, Defense Agency of Japan, *Nihon no boeki seisaku* (Japanese defense policy), 17 November 1978, p.1, or any of the annual white papers on defense published since 1975.
6. Defense Agency of Japan, *Defense of Japan, 1988*, p.75.
7. *Asahi Shimbun*, 20 June 1987; *Japan Times*, 21 June 1987.
8. *New York Times*, 30 August 1987.
9. See Kenneth B. Pyle, "Japan and the Twenty-First Century," in Takashi Inoguchi and Daniel I. Okimoto, eds., *The Political Economy of Japan: The Changing International Context* (Stanford: Stanford University Press, 1988), pp. 452–57.
10. Naikaku Sori Daijin Kambo Kohoshitsu (Office of the Secretariat to the Prime Minister), *Seron chosa nenkan, 1967* (Public opinion research yearbook, 1967), Tokyo, 1967, pp. 217-18.
11. Japan has agreed to defend the sealanes 1,000 miles around its territory but has been slow in implementing even this modest plan.
12. Henry Kissinger and Cyrus Vance, "Bipartisan Objectives for Foreign Policy," *Foreign Affairs* 67 (Summer 1988), p. 913.
13. See, for example, Mike Mansfield, "The U.S. and Japan: Sharing Our Destinies," *Foreign Affairs* 68 (Spring 1989), p. 9. This

article, especially the sections on defense and air, endorses in
extreme form all of the myths discussed here.

14. The Comprehensive National Security Study Group, *Report on Comprehensive National Security* (Tokyo: Prime Minister's Office, 1980), passim.

15. Nicholas Lardy, "Structural Change in the World Economy, Country and Regional Prognostications: China," Curry Foundation Conference, Washington D.C., 27 October 1988, passim.

FOUR

Contemporary United States–Korean Security Relations

Edward A. Olsen

United States–Korea Cooperation

Since the United States enlarged its commitment to the defense of South Korea during the Korean War, American forces have remained there to sustain it. Before examining specifics of that commitment it is worth putting those forces into the context of United States global strategy, of which they are part—albeit an outsize portion. This is important intrinsically and because that broad strategic context often is overshadowed by its Korean focus. The latter propensity is typical of Korean analysts. The United States has some 43,000 members of its armed forces serving in South Korea—approximately three-quarters of them in the army, one-quarter in the air force, and a small number in the navy and marine corps. Those numbers are significant because the American forces assigned to Korea remain remarkably large a third of a century after the hostilities there were formally halted by the 1953 truce. There have been three major efforts to scale down United States forces assigned to Korea: by President Eisenhower in the mid-1950s, by President Nixon in 1970, and by President Carter during 1977–1979. The first two succeeded, the last failed. The aggregate numbers also are important in the Pacific region because they are the only major deployment to a theater that approximates a combat zone.

The proportion of American forces from each of the services also is significant. Unlike the other Western Pacific locales of American bases (Japan, the Republic of the Philippines, Guam, and the Marianas), in which navy and air force personnel are a clear majority, Korea is essentially an army-run theater. This is not to suggest that other United States services (particularly the air force) do not play important roles in Korea, but the army is the dominant branch as reflected by its leadership of all major joint commands in Korea. This feature of the United States commitment to Korea is important in two respects. First, strategically, it reflects a significantly different United States strategic approach in the Korean theater than is evident in the rest of the Pacific. Most of this region is integrated into the United States' global strategy, with its focus on air, naval, and ground mobility. In contrast to the relatively fixed-place American commitments with NATO, continental in orientation, the Pacific-Indian Ocean area represents a region of flexible, offshore, maritime strategic responses. In the midst of all this, however, is the United States commitment to Korea, which strongly echoes the continentalism of the United States position in Europe. In these terms American forces in Korea, though part of the United States Pacific Command structure, operate in a more static and fixed environment. This is not cited as a criticism of these forces, or of their different roles, but to point out the differences.

However, growing out of these differences are aspects of the United States force presence that sometimes do warrant criticism. Because of service rivalries among the United States armed forces, some problems of intratheater cooperation do arise. These are exacerbated by friction over the proper role of forces in Korea in wider strategic affairs. Should United States forces in Korea be targeted solely against North Korea or be considered redeployable? Also aggravating problems associated with the roles of United States forces in Korea are the inertia of policies linked to a thirty-five-year-old, partially anachronistic system; the vested interests of the U.S. Army in perpetuating the near combat-level experiences provided by an assignment to many units in Korea; and the personal attachments to Korean defense interests of many senior officers who have served more than one tour of duty there. These factors induce a significant prejudice favoring existing policy and existing deployments and tend to skew the perspectives in United States military decisionmaking. These important elements must be borne in mind as we consider contemporary American security interests in Korea.[1]

The United States commitment to Korea of the 1980s and

beyond is based on the Mutual Defense Treaty between the United States and the Republic of Korea, signed in October 1953 and put into force in November 1954.[2] Though designed as a treaty between two utterly different states, one the richest and strongest power in the world and the other a war-wracked client state with few visible means of support, it has stood the test of time very well. As South Korea—thanks, in part, to its reliance on the United States treaty—has grown in economic, political, and military stature, the provisions of the treaty increasingly appear prescient. Though many Americans have considered ways in which United States–Korea security ties could be improved under the provisions of the treaty, few have seriously advocated jettisoning the pact and abandoning South Korea. The present analysis does not stray from that sound precedent but explores some innovative ways of bolstering bilateral security ties.

Even those Americans well known as critics of Seoul's authoritarian politics ordinarily are willing to acknowledge that a serious threat to South Korea exists and must be addressed and that the treaty plays a valuable role.[3] The support of Korea by the United States government is a given in the equation. Ever since the mid-1970s, when American credibility in Asia was adversely affected by severe reversals in Indochina, Washington has regularly assuaged the anxieties of Asian leaders by reiterating—often ad nauseam—the sincerity of United States commitments. That has been clear in South Korea, where leaders periodically flaunt their nervousness, eliciting an entirely predictable American reaffirmation of support for the Republic of Korea.

President Chun Doo Hwan's first, and ballyhooed, visit to Washington in February 1981 was a graphic example of these expressions of mutual support.[4] Even Chun's subsequent hyperbole in 1981 about South Korea's having "the means" to reduce North Korea "to ashes" (which was taken by some as a suggestion that South Korea possessed nuclear weapons) did not shake United States support.[5] Throughout the Reagan years, American support for its commitment to Korea remained strong, sometimes excessively so. In the wake of the Soviet Union's shooting down of KAL 007 in September 1982 and North Korea's 1983 mass assassination of sixteen prominent South Korean officials accompanying President Chun to Rangoon, a new level of anxiety arose in South Korea. There was speculation about the meaning of these assaults focusing on whether they were calculated tests of United States–Korea resilience. By 1988 those events appeared more attributable to internal decisionmaking in the Soviet Union and North Korea, not necessarily connected, and not as carefully calculated as initially feared. In any event, the United States response was both correct and speedy. Washington affirmed once more its commitments in Korea during the Reagan-Chun Seoul summit in early November 1983, just weeks after Chun had escaped assassination in Rangoon.

Such expressions of support when aroused by events beyond the control of the United States are serious enough, but they are made more serious when stimulated by the actions of Americans. That is precisely what happened in April 1985, prior to the third Reagan-Chun summit, which was slightly tarnished by tension surrounding United States pressures on trade and human rights policy, but which was seriously marred by Seoul's allegation that eighty-seven United States-built helicopters had been shipped to North Korea via sales to West Germany.[6] Those charges subsequently proved accurate and became a serious problem for United States–Korea defenses against the North. Most important is that episode's role in American escalation of material support for the Republic of Korea and its again underlining the durability of its commitment to South Korea.

As North Korea and the Soviet Union intensified their military cooperation during 1985–1987, the levels of United States support to the South followed suit. Apprehension over that cooperation and what it might mean in terms of North Korean willingness to run risks appears to have stimulated new levels of rhetorical and actual support. Secretary of Defense Caspar Weinberger, voicing United States concerns over North Korea's ability to disrupt the fall 1986 Asian Games and the 1988 Olympics, in April 1986 promised that the United States would help ensure the safety of both events and then told Korean Defense Minister Yi Ki Baek that Korean security is "pivotal to the peace and stability of Northeast Asia, which in turn is vital to the security of the United States."[7] United States Ambassador Richard L. "Dixie" Walker subsequently repeated the "vital" characterization in a Cheju-do speech before the Korean Management Association.[8]

The logic behind such assertions must be questioned. As diplomatic rhetoric it is understandable, but if taken seriously, as evidently intended, it simply does not make sense. While the United

States is justified in worrying as much as South Korea about the implications of closer North Korea–Soviet military ties, and should continue to caution Moscow and Pyongyang as the then-head of the United States Pacific Fleet (Admiral James "Ace" Lyons) did on a tour of the United States' Pacific allies,[9] Washington should not overreact. Even assuming any worst-case situation growing out of Soviet-North Korean military cooperation,[10] the only state with a truly vital interest at stake is the Republic of Korea. Its very existence is at risk. That is not true of the United States' stake in the issue. Should the worst occur and South Korea go the way of South Vietnam, the United States would still survive. To be sure, the United States would have lost a close ally, a valued trade partner, a key buffer in the defense of its other Northeast Asian treaty partner (Japan), much credibility, and—last but certainly not least—many friends. However, the United States would continue to exist, and many Americans would soon put such a reversal out of their minds—as quickly as they dismissed the Republic of (South) Vietnam from their consciousness. In this context, contemporary characterizations of American interests in South Korea as "vital" are not credible. This may seem a harsh judgment, but candor in these matters is essential. The United States and the Republic of Korea are better off being candid.

South Korea clearly has vital interests in the security commitments it enjoys from the United States. The reverse is not yet true and may never be true. That is not to denigrate the important United States security stake in the Republic of Korea or imply that it does not play a vital role in regional affairs, in which the United States also has very important interests. If one itemizes the threats in the region to which the United States and South Korea respond, there are major areas of overlap. These are so large that many analysts loosely refer to them as common, or shared, security interests. Clearly, they are larger in scope and depth than any so-called common interests shared by the United States and any other Asian friendly state or ally. Just as clearly, this makes South Korea a particularly compatible ally; yet it is nonetheless true that our security interests are not fully in harmony.

How are they similar and how do they differ? That is crucial to an understanding of contemporary United States–Republic of Korea security relations. The most basic issue concerns our respective threat perceptions. Both the United States and Korea are motivated by concerns about communist aggression. Specifically, they share concerns about another round of North Korean aggression. Each recognizes that North Korea is an armed camp whose leaders are driven by a mandate to impose Kim Il Sung's vision on all of Korea. Both recognize that North Korea's offensive troop deployments and tunnel digging are clear signs of its aggressive intentions, that North Korea is a revolutionary state anxious to export revolution and terrorism to distant corners of the world, and that the Soviet Union still may be a dangerous factor in Korean affairs because it might pursue policies that could produce or sanction North Korean violence against South Korea. Finally, both understand that neighboring China and Japan can play useful roles in coping with North Korea and/or the Soviet Union. This is a broad area of agreement.

However, the differences also are significant.[11] The potential communist aggression faced by the United States and the Republic of Korea are significantly different. Though both kinds are palpable and could inflict massive destruction on the American and South Korean people, the threats are markedly different. South Koreans live routinely with their North Korean enemy poised only a short distance away. Further, they believe the joint efforts of their own and United States armed forces are all that prevent North Korea from trying another assault should it—in the face of the countervailing logic we now hope will dissuade it—decide on such an action. However, war for South Koreans, while certain to be devastating to the progress they have achieved and costly in terms of lives, would not be likely to mean the utter destruction of the South Korean people and territory. Even if the Republic of Korea lost the conflict, most South Koreans would survive and live on under the thumb of a communist regime. Though that prospect is not welcome, and clearly South Koreans are ready to do their utmost to prevent it, such a war is realistically less terrifying and more bearable for South Koreans than the possible conflict confronting Americans. Few Americans visualize a superpower war as either winnable or survivable. Because of these very different perspectives on war, Americans and South Koreans hold very different threat perceptions and fears.

The United States also is concerned about North Korean aggression. Washington fears for the safety of its South Korean allies. To that extent the concerns are shared. However, Americans also are fearful of the impact of North Korean aggression on Japan and China, and of possible American inability to cope with their reactions to United States responses in Korea. Washington also is concerned with the effects of such interactions on United States interests in NATO, the Middle East, and globally. As a superpower, the United States must think in much broader and more complex terms about the North Korean threat than South Korea does. A famous cartoon during the Korean War depicted President Truman showing General MacArthur the Korean peninsula writ large on one plane of a cube-shaped globe, advising him that "there is more to the world than this." One consequence of this truism, sometimes still ignored in nationalistic and parochial Korean circles, is a very different readiness to consider the American military in South Korea as deployed combat forces as contrasted with instruments of deterrence. While another war in Korea would not automatically mean confrontation between United States and Soviet forces (as in a NATO-Warsaw Pact clash), the stake of the Soviet Union in North Korea is so great that Americans must contemplate the possibility. Consequently, South Koreans tend to accept the notion of actually using joint conventional deterrence more readily than do Americans.

The North Korean Threat

This difference in perspectives contributes to another normally unarticulated rationale for the United States presence in Korea, namely, to keep South Koreans from "going north." This was much more important in the past than it is now. Many contemporary analysts utterly discount this possibility. The common wisdom today holds that South Koreans would no longer be as willing as they once appeared to risk everything on an attack northward. Yet three corollary considerations make the common wisdom seem less wise. First is the possibility heard in some Korean circles that at some point the Republic of Korea will become unilaterally invincible compared to North Korea. Then conquering the North may again be appealing. Second, South Korean nationalists—civilians and military—clearly chafe under the constraints imposed by the dependent role persisting in United States–Korean security relations. If they ever had a chance to turn the tables on North

Korea and no longer felt they were on an American geopolitical leash, would they try? No one can be certain of the answer. Third, and most troubling, is the possibility that the Republic of Korea might one day opt to go nuclear to be fully independent, to replace a diminished level of United States support, or to keep up with the nuclear proliferation among its Asian neighbors. Against this background, the American commitment to Korea assumes several different sets of overtones to South Koreans and Americans.

Lastly, while South Koreans share with Americans a desire to see China and Japan play security roles that could assist the United States and South Korea in coping with North Korea, they do not see either country in ways that are congruent with United States perspectives. South Koreans (and North Koreans) have long historical memories that include a perception of China as the Middle Kingdom and of Japan as an expansionist state. Both Americans and South Koreans normally recognize that China and Japan today differ from their past geopolitical incarnations. However, South Koreans harbor doubts about the depth of their neighbors' new facades. Many South Koreans persist in viewing Japan as a sinister factor in their future, frequently visualizing Tokyo as thirsting to again dominate Korea economically, politically, and—if the opportunity ever presented itself—militarily. Though China is not so intimidating to South Koreans as Japan, they would prefer to prevent China from reasserting its traditional paternalistic role vis-à-vis Korea. Though North Korea's rhetoric toward Japan is more harsh and explicit than South Korea's, and Pyongyang's apprehensions about Chinese hegemony are necessarily constrained by its fraternal ties, North Koreans view these neighboring major powers as long-term security threats in ways akin to South Korean perspectives. In short, the United States and South Korea clearly do not share the same global perspective when assessing South Korea's neighbors in regional security. South Koreans cannot afford the luxury of forgetting the past and accepting the American perspective, which is rooted in the reality of the United States' Western Hemisphere location far across the Pacific.

This major difference in perspective has profound implications for the threat perceptions of the United States compared to the two Koreas regarding the Soviet Union. Pyongyang obviously does not see the Soviet Union as a "threat" in the sense we use here. Only in terms of North Korea's not wanting to permit Moscow undue influence over North Korean affairs can a Soviet "threat" be spoken of in the North Korean context. Most logical Americans would expect North Korean–Soviet security relations to be approximately what they really are; but extending that logic to South Korea could easily lead to the false assumption that the Republic of Korea sees the Soviets in ways that are very close to United States perspectives. Nothing could be more self-deceiving. Of course, South Korea does share the sense of danger still posed by the Soviet Union to United States–led global collective security. Not even contemplating the extreme of a civilization-ending nuclear war, in which everyone would lose (putting North and South Korea in identical camps for a change), a severe United States strategic setback at Soviet hands would be disastrous for contemporary South Korean security interests. Thus, when Washington calls upon its allies to be sensitive to United States warnings about Soviet military strength and seeks cooperation in reducing cold war tensions, it gets a genuinely attentive audience in South Korea. Furthermore, Seoul is not above using enthusiastic responses to United States appeals for support to reinforce the American commitment to allies who demonstrate steadfastness and empathy with the United States as it copes with the Soviet Union. Notwithstanding its legitimate and/or manipulative echoing of Washington's views regarding the Soviet Union, typically South Korean elites hold a much more localized and parochial attitude toward the Soviet threat potential to Korea.

There is little sense in South Korea that Soviet forces pose a direct threat. The Soviet Union is seen as a potentially dangerous supporter of North Korean aggression against the South, but that perspective is commonly offset by South Korean expectations that Moscow will inhibit irrational acts by Pyongyang so that peninsular instability does not force the Soviet Union into a confrontation with American backers of South Korea. Except for that relationship, which is as reassuring as it is threatening to many South Koreans, it is difficult for Koreans to visualize how the Soviets could pose a direct threat to their security. Moreover, the Republic of Korea tends to relegate the Soviet threat notion so overwhelmingly to the purview of Americans that Seoul has difficulty visualizing how South Korea could do anything in response to Soviet actions. In this context, and in light of a broader and more optimistic South Korean agenda regarding the Soviets, Seoul does not now rank concern about the Soviet Union high on its list of security priorities. Consequently, there has been a marked difference in threat perceptions between the United States and Korea, which is narrowing somewhat as cold war tensions diminish.

It is important to note that these significant differences do not materially impede sound United States–South Korea security relations, though this may change. For now, the factors contributing to commonalities are emphasized by both Washington and Seoul. The divergence of perspectives does not normally come into view because each side of the bilateral relationship is preoccupied with fulfilling its portion of the security burden. Seoul is too busy coping with North Korea to seriously consider putting more on its plate. Washington tends to appreciate what South Korea is doing on its own behalf as a valuable way of contributing to regional security. Anything an ally does to tie down the forces of the Soviet Union or its allies is one less thing Washington has to contemplate undertaking. This, too, may change in the future if the Korean cold war also thaws, but for the time being there is little impetus in either Seoul or Washington to pursue such change. On balance, then, the United States–Korea security relationship is a rather complacent one, with frequent reiteration by both American and Korean leaders that relations have never been better, they will continue to improve, and everything is under control. Such statements must be chalked up as boilerplate designed to stop the remnant advocates of the Vietnam syndrome and other naysayers from disrupting a generally viable security relationship between the United States and South Korea.

Nonetheless, there are doubts among South Koreans and Americans that belie the rosy rhetoric so often heard. The greatest area of continuing concern is the durability of the United States security commitment. Though the Reagan administration pointedly shelved the Carter administration's aborted troop cutback proposals, did not permit recurrence of the confusion surrounding the 1979 remarks allegedly made (and then denied) by Ambassador to Japan Mike Mansfield about Korea's still being outside the United States' defense perimeter,[12] and supposedly discarded the "swing strategy" that would redeploy United States forces from Asia to Europe, South Korean fears and uneasiness derive from a sense that American cunning, duplicity, or stupidity will one day produce a sudden abandonment of its commitment. Efforts by

Congress to reduce nearly all overseas troop deployments send chills down the spines of South Korean leaders. So far, such sweeping measures have stopped short of Korea, perhaps because Korea is recognized as a tension zone, and acknowledging that the Republic of Korea contributes more to its own defense, in money and manpower, than many less cooperative allies do. This recognition even moved liberal advocates of global troop cutbacks to veer out of their way to exclude South Korea, as Representative Pat Schroeder did in 1986.[13] Whenever congressmen have advocated moderate United States troop cutbacks in Korea, their legislation has routinely been soundly defeated. Just such an amendment to the 1988 defense budget was rejected in 1987 by the House of Representatives, 311–64. Nevertheless, in August 1988 the House Armed Services Committee issued a report calling on allies worldwide to contribute more to burden-sharing. Clearly, this is an issue that will not be silenced.[14]

On balance, such indicators are reassuring to South Koreans, but the occasional poll demonstrating that American popular support for the United States commitment to Korea might waver should North Korea actually launch another attack still causes unease in Seoul. Such unease was aroused by a March 1987 poll by the Chicago Council on Foreign Relations, which showed weak support for United States assistance to the Republic of Korea in the event of war, even as public support for NATO and Japan was growing.[15] Periodic recurrences of such perceived weakness in the United States commitment wounds United States–Korea security relations, already affected by the United States' loss in Vietnam. That issue is constantly aggravated, and American officials find themselves more than a decade after that trauma still having to reassure Seoul. Even though the Reagan administration enjoyed substantially more credibility in Seoul than its three predecessors, it too had to engage in what American officials consider "stroking" their nervous Korean counterparts. United States authorities at various levels bend over backwards to calm Seoul by reemphasizing what Americans assume to be a given. In addition to the Reagan-Chun summitry, prominent examples of frequent stroking include Secretary of State George Shultz's mid-winter 1983 reassurances at the Korean DMZ; Secretary of Defense Caspar Weinberger's 1982, 1984, and 1986 reassurances in Seoul at cabinet-level defense talks; and Assistant Secretary of Defense for

International Security Affairs Richard Armitage's strong support in February 1987 for substantial bolstering of the United States–Korea defense relationship.[16]

American encouragement to Seoul goes far beyond words. During the 1980s, a number of major United States arms transfers and sales to South Korea occurred. These enabled the Republic of Korea to quantitatively and qualitatively improve their forces' capabilities, part of the "FIP" (Force Improvement Plan) measures begun in response to initial post-Vietnam qualms about United States support. By far the most controversial and significant episode of this sort concerned Washington's policy reversal that fostered the sale of F-16s to Korea.[17] Though this strengthened South Korea, it also raised the ante in the North Korea–South Korea arms race and may have exacerbated the subsequent expansion in Soviet military aid to North Korea. Since then, the annual FIPs have strengthened South Korean forces, aided greatly by rapid expansion of domestic arms production capabilities. Despite that, Seoul rarely lets an opportunity pass to underscore to the United States that more needs to be done. For example, during the 1987 United States–Korea Team Spirit military exercises, President Chun visited the command post and made a political pitch for more American support for FIP.[18]

United States reassurances to Seoul also extend to moves to bolster the troop presence in Korea. Midway though the Reagan years, the United States commander of the Combined Forces Command (CFC), General Robert W. Sennewald, described the ways in which United States forces had been augmented by new troop deployments, thus strongly signalling the full rejection of Carter-era plans for cutbacks.[19] Other signals of a strengthened American commitment include the 1986 decision to deploy major updated weapons systems like the medium-range Lance missiles to replace the obsolescent Honest John and Sergeant missiles previously in the United States inventory in Korea. The potential nuclear capability of the Lance was evident to South Korean observers.[20] Such factors, backed by the Reagan administration's engendering of a more hawkish stance among the American public, produced the results desired by the administration, namely, displays of Korean confidence in the United States. Former Korean Ambassador Kim Kyung Won exemplified this confidence when he stated in early 1987 that he sensed a "qualitative change" in the appreciation of Americans for South Korea and noted new depth and bipartisan support for the United States commitment to Korea.[21]

This improved security atmosphere, from the perspectives of both Seoul and Washington, yielded significant results at the Nineteenth Annual US-ROK Security Consultative Meeting (SCM) in May 1987. Three measures emerged from those talks, touted by some as the start of a new era in United States–Korea security: an agreement on a defense hotline, a major increase in South Korean support for American forces stationed in Korea, and United States recognition of the importance of Korean defense industries.[22] Of the three, the two most important are the agreements on support and defense industries, because both denote areas of major change in South Korea, with tremendous implications for its future.

The decision to upgrade Korean support for American forces appears to have been the result of a Defense Department audit of the CFC that found the United States was "paying a disproportionate share" of running the CFC. The audit, conducted two years earlier, was not released until after the Nineteenth SCM in 1987.[23] These actions and Seoul's responses clearly reflected widespread recognition among Americans dealing with Korea that an increasingly prosperous Korea was not contributing a proportionate share to its own defense. Seoul is adamant that it pays its own way and likes to compare its defense expenditures (measured as a percent of its GNP and share of its national budget) to those of other United States allies, especially Japan. While those comparative proportions are impressive, they do not detract from the American view that the Korean defenses are heavily subsidized by the United States and that the demands on the Korean GNP and budget would be far greater if the United States were not paying and doing so much on South Korea's behalf.

Seoul's response to these pressures to pay more is embodied in the acronym "CDIP" (Combined Defense Improvement Plan), by which South Korea will contribute to the upkeep of United States forces in Korea. This plan has been around for some time and is used primarily as an instrument to supply American forces with better housing. This role is treated gingerly by Seoul, which fears that the Korean masses would not understand why their government provides much better housing for soldiers from a rich foreign country than that possessed by most Koreans. Far from wanting to help pay for its presence in Korea, some Korean politicans want the United States to pay "rent" in Korea, as some say it does in the Philippines.[24] Actually that disparity also rankles many Korean officials in private even as they acknowledge the political need for it and console themselves with the reality that Korean firms profit from the construction contracts and that eventually the projects built under CDIP would revert to Korea, when American forces in Korea are no longer necessary. Equally important, in time Koreans may understand how much they benefit financially from these arrangements. Were the United States to withdraw its forces and commitment from Korea, the defense-related tax burden for South Koreans would mushroom dramatically. Though paying more than they think is fair, South Koreans are getting a very sound deal, though one that some Americans dislike.

Perhaps most important, CDIP probably is best seen as a useful precedent, enabling Americans to press for more and new ways in which South Koreans can use their economic well-being to contribute to joint security efforts. While visiting Seoul in January 1987, former Ambassador to Korea William Gleysteen cautiously suggested the possibility that Korea could contribute much more to defense because of its economic progress.[25] Others are likely to be less circumspect in the future as the Korean economy continues to expand.[26] However, it is notable that the Republic of Korea seems to be readying itself for a broader security role. For example, though emphasizing North Korea as the primary threat at sea as well as on land, in his 1987 commencement address to the Republic of Korea Naval Academy President Chun also stressed the vital nature of sealane defense to South Korean interests and the dangers to those sealanes posed by the Soviet naval buildup in the Pacific.[27] This may not seem like much progress, but it is a major improvement over the parochialism in Korean defense circles that allows them to indulge in tunnel vision regarding the North Korean threat.

While the CDIP burden-sharing arrangement has economic implications in terms of cost, those implications pale when compared to the American acknowledgment of the importance of Korean defense industries. That sector of the South Korean economy received a major boost following American setbacks in Vietnam, when the Park government decided it would be imprudent to rely as heavily as before on United States arms suppliers. South

Korea faced two choices then—to strive for greater self-sufficiency in weaponry, or to seek other than United States sources. Though some diversification was begun, and continues, the other-source option was rejected as Seoul's main recourse for three reasons. It would have substituted one dependency for another, sent the wrong signal to Americans about Korean desires for material independence from the United States, and done little to help the South Korean economy. Seoul opted for a safe middle path, relying on American suppliers and also building major defense industries, but achieving this by co-production schemes with American arms producers. This approach enabled the Koreans to have their cake and eat it too, because the new levels of self-sufficiency did not undermine existing security relations with the United States.

All went reasonably well for Seoul until the domestic arms producers grew and began competing with United States arms producers. The technology transfer arrangements imposed by the United States supposedly precluded Korean sales to third parties that would compete with American producers. Despite the existence of such terms, Korean firms have antagonized some of their American counterparts, who see South Koreans as up-and-coming rivals who have proven more flexible in meeting customers' needs. In part because competent Korean arms producers emerged as a reality and because Seoul's new defense industrial capacities allowed it to threaten diversification from a new position of strength that would limit United States restrictions,[28] the United States eventually—albeit reluctantly—accepted South Korean arguments that the Korean defense industries were important to American security, complementing areas where the United States was weak. That argument was most visible in a cogent study done for the Heritage Foundation by a visiting fellow from South Korea, who happens to be a South Korean army colonel.[29] The double standard evident here of a Korean official, with vested interests in seeing his country's defense industrial base grow, writing an advocacy piece on behalf of an American research group with easy access to Reagan administration insiders, does not seem to have troubled many people.

An even more ironic aspect of the situation is that the complementarity that South Koreans sought to establish was only possible because of weaknesses in the United States defense industry base, which was in part created by foreign competitors like

South Korea! As the rust belts of the United States testify, and the debate over "deindustrialization," "hollowing out," and industrial policy demonstrate, the United States has lost substantial assets in heavy industry that are a key part of the foundation of the United States' defense industry. This, too, seems lost on many supporters of closer Korean-American defense industry cooperation. In any event, by 1986 the United States and South Korea at the Eighteenth US-ROK SCM had reached a preliminary agreement to create a US-ROK Defense Industry Cooperation Association, in which the United States delegation described South Korea's defense industry as a "priceless asset for the free world."[30] A year later, at the Nineteenth SCM, more definitive American acclaim for the competition was put on the record. The annual meetings of the private Council on US-Korea Security Studies also provide a supportive forum for industrial cooperation.

Against the background of such arduously developed rapport, one hesitates to point out the sore spots in ongoing United States–Korea security relations. Nevertheless, airing our troubles is essential to a better understanding of how the United States and South Korea differ. Many problems remain. They shall be categorized here into four major and several lesser groups. The "lesser" issues were not truly minor items but were of a transitory nature, whereas the major issues promise to be more lasting. One of the most inflammatory episodes involved the sale of eighty-seven United States–made helicopters to North Korea in 1983–84. Not only does this still cause concern in Korea because of continuing South Korean worries about North Korean use of this American-style equipment to engage in unconventional warfare, but in 1987 Seoul still professed not to understand United States explanations about the sale and expressed lingering concern that such a sale could not have occurred without high-level United States knowledge of it.[31]

Another type of short-term flap stems from exposures of various weaknesses in United States–Korean security relations. Perhaps the most significant of that genre was a 1984 House Appropriations Subcommittee on Defense report that scored United States military readiness worldwide, but was devastating in regard to Korea because it quoted a senior U.S. Army commander in the Pacific that if war in Europe coincided with war in Korea, "The US troops left in Korea had best learn to swim to Japan as there is nothing left to reinforce them." The Pentagon excoriated that report as "dangerous and wrong."[32] The controversy was contained by force of will in Seoul and Washington because neither really wanted to publicly delve too deeply into the specifics. The concerns cited in that report were real then, remain real today, and are still being remedied. They are a lingering sore point in United States–Korean security ties because of Korean fears that in a multifront United States war, a Korean theater would not be assigned top priority by Washington. Moreover, the evident continuing strains on United States defense resources since the report was issued—a real issue in the 1988 United States presidential campaign—cannot be reassuring to Seoul. Neither can Seoul be consoled by Washington's not providing swimming lessons for its forces in Korea.

Sometimes even the best-intentioned bilateral security moves can backfire. A classic case was Ambassador Edward L. Rowny's invitation to the Koreans to participate in SDI research with the United States, largely to assuage Korean anxiety that they were being left behind in United States–Japan cooperation on SDI and that the South Korean economy would suffer as a consequence of losing access to high technology. In any event, Rowny (President Reagan's adviser on arms control) spoke in Seoul on the SDI and America's Asian allies.[33] Though Korean press coverage of his visit was upbeat, many Koreans who heard his talk were affronted by Rowny's reported lack of interest in what South Korea might actually do to help and the suggestion that the United States was consulting with Koreans only as a courtesy. Instead of reassuring these Koreans, the episode underscored their concern that the United States and Japan would proceed in tandem, leaving Korea in their wake.

Another well-intentioned move involved the efforts of retired U.S. Army General John Singlaub to obtain South Korean assistance for the United States' support of anti-Sandinista "freedom fighters" in Central America. This activity was denounced by North Korea before it was widely publicized in the United States in the Iran/Contra hearings, where South Korea attained some notoriety by being one of the unnamed but numbered Asian countries.[34] Seoul was extraordinarily lucky that this fiasco in American foreign policymaking did not blow up in its face. Seoul had let its anticommunist fervor and willingness to cooperate with arch-right wing groups—such as Singlaub's World Anti-Communist League—

draw the Republic of Korea into a volatile arrangement. Seoul's desire to help the United States in a global cause cannot be faulted, but its connivance in a movement to subvert American legal processes strongly echoes its actions in the Koreagate episode. Should further instances of this sort come to light, South Korea's readiness either to manipulate the United States government directly or to help American manipulators of that system could again prove disastrous.

If South Korean reactions to events in the United States can be disruptive, so, too, can American reactions to events in Korea. By far the most disruptive was the fiasco surrounding Seoul's gross mishandling of the putative "death" of Kim Il Sung in November 1986. Evidently created by rumors, sloppy South Korean intelligence, poor command communications, and a Korean defense minister—Yi Ki Baek—who evidently got carried away in his enthusiasm, this farcical episode badly damaged Seoul's international credibility. South Korea cannot afford to be caught crying wolf in any circumstances concerning North Korea. Making matters worse, a poorly concealed behind-the-scenes controversy broke out between the United States and South Korea because of Seoul's repeated efforts to shift responsibility for the fiasco to Americans on the scene. This incident was almost a caricature of the poor communications and frictions that can develop between South Koreans and Americans in the CFC, except that this was a real portrait.[35]

Since late 1986, South Korea has been calling attention to North Korea's construction of its Mt. Kumgang Dam, upriver from heavily populated areas of South Korea, as an extreme security threat. In response, Seoul has pushed a counteractive dam known as the Peace Dam. Seoul raised nearly a third of the cost of this defensive dam through public donations.[36] American reaction to the new threat, and to Seoul's hyperactive response, was lukewarm at best. Many American officials could barely conceal their private chagrin at Seoul's overreaction and its use of melodramatic telethon fundraising techniques. This difference of opinion did not cause waves in the United States, but it ruffled Korean feelings.

Two other American reactions to Korean security arrangements also made little impact in the United States but were felt in Korea. One was the highly negative critique of the U.S. Second Infantry Division's role in Korean security by an experienced United States officer, Brigadier General John C. "Doc" Bahnsen, in an influential military journal (*Armed Forces Journal International*). Koreans can discount amateur carping about the United States commitment to Korea, but this professional critique—which described the Second ID as an anachronism that should be disbanded, is kept in Korea to appease political needs, and is doing a job that South Koreans should be doing for themselves—could not be rebutted. Instead, it was ignored whenever possible by South Koreans. Though General Bahnsen's critique is even more valid today, it has not received much attention in the United States, and South Korean officials hope that that will not change.

The other negative American reaction to a Korean security development concerns the mid-1987 public announcement by Seoul that the Korean navy had a domestically made submarine.[38] Washington had long opposed South Korean desires to build and deploy a submarine to match those in the North Korean naval inventory. American logic held that South Korea needed better antisubmarine warfare (ASW) capabilities, not its own subs. Whether for strategic reasons or from a desire to keep up with the northern Kims, Seoul followed its own lead and did what it wanted. Regardless of what else this development may indicate about United States–Korean cooperation, it clearly shows Seoul's stubbornness and willingness to chart a different course from that prescribed for South Korea by Washington. American reactions to this development have been pointedly understated and cool, suggesting unhappiness about the decision but resignation to a fait accompli.

Of the four major areas of bilateral divergence, the easiest to manage is the nuclear issue—normally concealed behind a screen of official secrecy. The United States government's standing policy is never to confirm or deny the existence of, or plans for, American nuclear weapons anywhere in Asia. This is clearly enunciated regarding Korea. The Korean government echoes that policy, though distinctly less forcefully. Despite this aura of secrecy, the presence of American nuclear weapons in Korea is widely assumed in the United States and Korean (South and North) media. Normally, discussion of that factor in American strategy toward Korea is prudently couched in terms of a United States nuclear umbrella shielding South Korea and constituting a vital portion of the deterrence equation that keeps North Korea at bay. That formulation will be used here, too, for most of the discussion on Korean deterrence. However, occasionally the United States and Korean media do make specific accusations about American nuclear deployments in Korea that cause considerable unrest. Two prominent examples of this affected United States–Korean security and produced evidence of very different American and Korean reactions to the possible use of nuclear weapons in Korea. In May 1983, columnist Jack Anderson reported the existence of twenty-one ADMs (Atomic Demolition Munitions), or nuclear landmines, south of the DMZ and cited Pentagon interest in using neutron bombs should another Korean war occur.[39] Anderson's claims were repeated in June 1984 and caused consternation in Seoul.[40] In April 1985 the authoritative *Bulletin of the Atomic Scientists* also alleged twenty-one ADMs were deployed in Korea.[41] Both the South and North Korean media reported this assertion.[42] North Korea was apoplectic about these weapons allegedly being poised to defend against attack, but South Korea also showed apprehension that such devices were being discussed openly. North Korea routinely alleges that the United States is making South Korea a "forward nuclear base" to threaten all of Asia and has described extensive scenarios calling for use of United States theater nuclear weapons in Korea.[43]

South Korean leaders are unimpressed by North Korean accusations about the United States because they contribute to North Korean fear, which helps create the deterrence that makes South Korean security viable. However, two nuclear issues do cause concern in Seoul. One is the notion that South Korea might pursue its own nuclear option. Though often ranked among those states with the scientific and material capability to "go nuclear," Seoul regularly disavows any such intentions. Following Jack Anderson's assertions, the Seoul government went so far as to spell out publicly its reasoning that the United States nuclear umbrella obviated an independent Korean nuclear capability and that South Korea would not alter its rejection of that option so long as the United States nuclear shield protecting Korea remained in place.[44] Framing the relationship in this manner inadvertently revealed one of the most sensitive areas of United States–Korean security differences. In effect, the American nuclear umbrella is held hostage by South Korea's implicit threat to go nuclear if the United States wavers and withdraws its protection. A rarely discussed corollary to this hostage relationship is the de facto ability of Korean armed forces to wield ultimate control over any United States nuclear weapons that might be in or transit Korea. It is doubtful that the United States is truly free to close its nuclear umbrella and go home for fear that South Koreans would either seize the "umbrella" or would replace it with one of their own. Either result would have horrifying implications for United States policy in the region and for nuclear proliferation.

The second aspect of the nuclear issue that concerns Seoul and causes differences in the bilateral security relations, is Washington's possible intention to use nuclear weapons in retaliation against any North Korean aggression. Seoul is ambivalent on this issue. When South Koreans (or other allies) hear prominent American defense specialists question willingness to sacrifice an American city for an allied city in a nuclear exchange, the allies get extremely nervous. Stansfield Turner in 1986 called the nuclear umbrella protecting Asian allies a "myth."[45] South Korean leaders are always ready to hear affirmations from United States officials that the "myth" itself is mythical. While Seoul welcomes such reassurances, sometimes the reported willingness of American defense officials to use nuclear weapons againt North Korea seems a bit too enthusiastic. The South Korean media has cautiously expressed concern that the United States might actually resort to using nuclear weapons in Korea. The *Chosun Ilbo* editorialized in early 1983, "We shudder with dread over a nuclear war [in Korea]—we are seized with panic and uneasiness because of nuclear damage even if we win this war."[46] Behind the concern lies a major queasiness in the South Korean psyche. Ever since Hiroshima and Nagasaki, the Japanese argument that white American leaders were racially motivated in their readiness to use nuclear weapons against Asians but not Europeans has found a receptive audience among many Asians. It surfaced in the heavily controlled South Korean press in 1984, indicating a sanctioning of these fears by South Korean officials.[47] It is easy for rumors of United States proclivities to use nuclear weapons in Korea rather than against Warsaw Pact adversaries, for example, to attain credibility among South Korean elites and to filter down to the public. It should, therefore, be no surprise that some tensions exist between Americans and South Koreans over the under- or over-willingness of the United States to use nuclear weapons in Korea. These different perceptions, and their subtle influences on both sides, raise uncertainties in this aspect of the security relationship. This facet of bilateral tensions is normally shielded from public scrutiny by each country's reluctance to talk openly of such matters.

The second of the four major problem areas in the relationship involves the respective roles of United States and Republic of Korea forces in the Combined Forces Command system. There are three troubled aspects of this system. One is important for tactical joint operations in Korea, namely, the many routine problems that develop when two states as culturally and materially different as the United States and the Republic of Korea must work as a team. These problems can be solved by sufficient attention.[48] Far more serious are the other two aspects, opposite sides of one issue: the influence of United States forces in Korean affairs. Many United States and South Korean critics allege that—despite official denials—the American emphasis on strategic factors in Korea has led to the militarization of the Republic of Korea.[49] This stems from accusations that the United States tolerates, sanctions, or enforces a military dictatorship in Seoul with related problems in political and economic affairs. These matters cause friction in bilateral security relations because American and South Korean defense officials are sensitive to, and inhibited by, the accusations. A counterargument, which holds that the United States military does not enjoy enough influence over, or respect from,[50] their Republic of Korea counterparts also is plausible. The latter position supports those who believe the United States should use its power to exert influence on South Korean leaders in order to nudge them toward political and/or economic reforms.

United States–Korea frictions in the CFC are poorly disguised. They are evident in the arrogance of each side toward the other. American arrogance usually is predicated on big-power chauvinism, while South Korean arrogance is based on a complicated mixture of confidence in the superiority of Korean martial values, hubris deriving from Korea's ancient and homogeneous cultural and ethnic background, and defensive anti-foreign sentiments stimulated by often-overbearing Americans. The structure of the CFC, with an American general in charge, clearly grates on Korean sensibilities. Korean officers often do not successfully hide their feelings that they should be running their own security. Since South Korea's defense is provided largely by Korean forces, it is easy to sympathize with these rather nationalistic officers who chafe at being underlings in an American-led hierarchy.

The criticism leveled at the United States for perpetuating a "puppet" and a "dictatorship" by manipulation of the Korean military is difficult for responsible figures in the South Korean armed forces to stomach. They know in their hearts that this popular impression is false and say so, but to little avail. Americans, too, try to correct these misimpressions of United States dominance, but because of the CFC structure they also are usually unpersuasive.[51] Both Seoul and Washington seem reconciled to American leadership of the CFC as a lesser evil. This is because it is necessary for both sides to smoothly connect the United States–led Western security network with integrated South Korean defenses, to sustain the linkage of United States forces in Korea with their United States hierarchies, and to maintain the logistical pipeline to resupply the defenders of South Korea. Less often cited, but no less important, is the need to retain a United States commander in a hierarchy with jurisdiction over nuclear security issues. For these reasons, neither Washington nor Seoul now appears ready to change the CFC. To resolve some of the problems, though, there may be reasons to make some changes.[52]

Korea, Japan, and Northeast Asian Security

The most sensitive area in Korean-American security relations concerns a factor that officially does not exist, that neither side supposedly wants, and that may never materialize. This is the potential role of Japan in Northeast Asian regional security and, perhaps, in Korean security. A central problem in American security policy toward Korea has long been the role of Japan and how to explain that role to Koreans. The United States' early strategy toward Korea was shaped and reshaped with one eye firmly on the geopolitical significance of events in Korea for Japan, and the United States' stake in that country. Since the Korean War, the United States has purposely maintained a distinction between its security policy toward Japan and that toward Korea. This is an artificial division, for the geopolitical significance of each state to the United States has been used to justify American commitments to the other. Moreover, the fundamental threat to each ally comes from the same source, the Soviet Union, and the United States' perception of that threat is region-wide, not arbitrarily demarcated in the center of the Sea of Japan. The only reason the United States maintains a partition between United States–Japan and United States–Korea security relations is that Tokyo and Seoul are adamant that no formal linkages should be permitted because of Japanese-Korean historical enmity and contemporary frictions.

This is the setting for the potential role of Japan in regional and/or Korean security. The logic is problematic: United States–Japan security disagreements have been muffled, especially by the Nakasone administration's good relations with the Reagan administration. Although limited but real progress has been made in United States–Japan defense burden-sharing, the appearance of supposedly major gains is illusory. The huge inequities in bilateral security ties remain—with the United States bearing the lion's share of the costs, risks, and responsibilities—ready to further sour a relationship already seriously strained by massive United States–Japan trade frictions. The linkages between these troubled facets of relations between the two countries are latent now, but they are a potential source of serious divisiveness. United States and Japanese perspectives on defense and trade are not nearly as harmonious as Washington and Tokyo would like their respective publics to believe.[53] These differences and the potential for serious disruption of the relationship were highlighted by United States reaction to the Toshiba violation of COCOM restrictions on exports to the Soviet Union, which precipitated a strong American response with clear repercussions for trade legislation in 1987 and 1988.[54] There is an excellent chance that disagreements will become even more important in international affairs now that both Nakasone and Reagan are out of office. Neither the Takeshita nor Kaifu administration is coping as well as Nakasone's did.[55] It remains to be seen how well the Bush administration will deal with the challenge from Japan. NATO is on the edge of what many consider "de-coupling" from the United States, the product of renewed willingness of Western Europe to ask, as one prominent Frenchman did, "Do you think 320 million Europeans can continue forever to ask 240 million Americans to defend us against 280 million Soviets?"[56] Asia is on the verge of facing an even more daunting question. Why do Asia's masses, particularly in its wealthy states, require levels of United States strategic support devised to meet the needs of poor postwar Asia? Why should the United States expend $47 billion on the defense of the Pacific while Asian states pay relatively little? Precisely how much more South Korea, Japan, or other allies should pay—or what sort of new burdens they should assume—is a matter to be determined by the United States in consultation with those allies, but whatever arrangements emerge should be far more equitable, or the United States may well reject them.

The existence of this profound question and the potential answer to it are concealed more effectively in Asia than in Europe. Nevertheless, the question causes great anxiety in Tokyo and Seoul, where officials sense the handwriting on the wall without actually seeing it. Because of both countries' clear desires that the United States not broach the notion of a Japanese role to defend anything beyond the Japanese islands, Washington has scrupulously adhered to a carefully crafted two-pronged strategic posture in which each Asian partner is urged to do more for itself, but the United States has shied away from the obvious linkages.

Despite near unanimity on this issue in official Washington, there have been a few deviations. In 1982, then Deputy Secretary of Defense for Policy and retired Army General Richard G. Stilwell called on Japan to engage in collective security by assisting with the defense of South Korea.[57] More recently, the House National Defense Task Force, composed of Democrats, in 1986 issued a report saying, in part, "In the long run, we favor a much greater contribution from Japan to the defense of South Korea." This wording caused considerable alarm in Seoul.[58] In practice, however, the closest the United States has come to suggesting a move of this sort were the mid-1980s requests that the Japanese Self-Defense Forces join in maneuvers with United States forces stationed in Korea.[59] This may seem a major step, but it is only an incremental change from the joint exercises held between United States forces stationed in Japan and the SDF. All those exercises focus on the defense of Japan, not on defense of the United States, and do not involve United States forces in Asia that are not actively defending Japan.

Two different levels of Japan-South Korea military cooperation already take place. One is relatively well known. Beginning in the late 1970s, ex-uniformed military or ex-defense bureaucrats from Japan and the Republic of Korea paid courtesy calls in, or made informational tours of, each other's country. Some were privately sponsored by right-wing organizations in each country and had clear geopolitical overtones, while many other tours were symbolic in their bureaucratic ordinariness. These were followed by exchanges of active-duty bureaucrats, uniformed defense personnel, and politicians on government defense committees.[60] Though these exchange visits are interpreted widely (especially in North Korea) as precedent setting in creating a triangular defense relationship, they are thin evidence at best. The second kind of Japan-Korea security cooperation is much less visible, but potentially more precedent setting. Since the Japanese Air SDF and the Korean air force flight zones abut each other and the Maritime SDF and South Korean navy sail in the same international waters, these branches of each country's armed forces are compelled to interact with each other and United States forces so that they do not get in each other's way when engaged in routine operations or exercises. Such so-called "PASS-EX" cooperation is inevitable and certainly invaluable, though not necessarily intended as a form of preliminary interaction among the forces, using equipment that is largely interoperable. As Japan and Korea operate further afield and engage in larger exercises with the United States, a continuation of this "accidental" form of cooperation can be anticipated.

Japan, Korea, and the United States all seem content to leave trilateral relations in place as they are. In Korea, the status quo inertia is supported by somewhat conflicting motives. There are significant voices that argue for upgrading trilateral security relations, but they appear motivated by two uniquely Korean incentives. Many South Korean advocates of this cause share with American critics of Japan a desire to see Japan spend more money on defense. There are many South Koreans as well as Americans who see Japan as a "free rider" benefiting not only from United States strategic largesse but also from the contributions to Japanese and regional security made by the Korean armed forces in defending a peninsula that many conservative Japanese see as a potential threat to Japan. This logic is persuasive to Americans and is often used by Seoul. However, these same South Koreans routinely insist that the United States not let Japan rearm too much and become a threat again.[61] The other common motive among South Korean advocates of trilateralism is a desire to elevate Korea to equality with Japan in the relationship. Many researchers in Republic of

Korea government-affiliated think tanks urge such equality in meetings with Americans. Increasingly, this quest for some sort of parity has been integrated into the Republic of Korea's policy goals vis-à-vis the United States, and Washington seems receptive. If the trend continues, it could have major implications for the future of United States–Korea relations. It is interesting to examine why Koreans are pressing this case.

One of the givens of Japan-Korea relations regarding the interjection of the United States as a security benefactor is the virtually identical attitude of each side toward Americans. Each is certain that it knows the other far better than any American can ever hope to know Korea or Japan. While there is some truth to such a belief, neither side recognizes that Americans can be more objective than either of them because of their antagonistic historical legacy. In United States–Korean security relations, Seoul officials are often smugly sanctimonious in their confidence that their ties with the United States are *the* core relationship and that United States–Japan ties should be ancillary. The long-standing American preoccupation with Japan nearly always is seen in South Korea as incredibly shortsighted and muddleheaded. This is an issue on which South Koreans of nearly all political stripes can agree. For example, in 1985 the main opposition party—which often criticized the Japanese economic presence in the Republic of Korea—argued similarly in favor of Japan's making "a positive contribution toward promoting security in the Far East," of reducing the United States' "lopsided" treatment favoring Japan, and for making "the security of Korea the pillar of peace and security in the Far East."[62] There has been no significant difference on these three issues between the ruling and opposition parties.

If there is a key element in South Korean opinion about Japan's role with the United States in regional security, it is Seoul's fear of that role escaping United States control and for the Republic of Korea to lose influence in American "control" of Japan. This prospect raises, for many contemporary Koreans, the specter of a resurgent, militarized Japan out to reconquer Asia. Though this is utterly unrealistic in contemporary circumstances, South Koreans still worry about the possibility. Considering that, and North Korean paranoia,[63] South Korean officials have been unequivocal in their rejection of trilateralism. Perhaps most critical was Foreign Minister Lee Won Kyung, who flatly rejected the notion several times, calling it "undesirable," "not worthy of study," and not compatible with the "two axes" of United States–Korea and United States–Japan security ties.[64]

Despite that broad anxiety and inhibitions against building upon the limited precedents already established, there is one offshore realm in which Korea and Japan increasingly agree about their security: protection of the sealines of communication (SLOCs). As a result of United States espousal of freedom of the seas in the Persian Gulf during 1987, attention was focused on the importance of shipping lanes for oil supplies crucial to the European allies and Japan. Those allies were widely criticized for being slow to react to threats to their own oil lifelines. Normally overlooked in that scenario was the importance to South Korea of Middle East oil and shipping, though the oversight is understandable because of the relative scale of the Korean economy compared to those of Western Europe and Japan.

Seoul has been slow to recognize the importance of SLOCs to its security despite several Korean ships' having been attacked in the Gulf, starting in 1982.[65] The Republic of Korea is a major contractor in the Middle East, and many South Korean citizens serve as crew members on Korean ships and those sailing under other nations' flags. Seoul's early response to troubles in the region was to offer various appeasement policies. Its coping mechanism in the region was a close copy of Japan's response, not the United States'. Neither Japan nor Korea considered itself capable of playing an active role in regional peacekeeping, though each briefly considered dispatching forces in a UN-sponsored peacekeeping mission. That was rejected by Seoul because of continuing concern about North Korea and for fear that sending Korean forces out of Asia would reinforce the arguments of American critics who want to remove United States forces from Korea on the grounds that they are not necessary if South Korea can spare its forces for duties abroad. Japan rejected the notion on the familiar grounds that it would violate its constitution.

Partly because of United States pressure on Japan throughout the 1980s to enhance its capabilities and commitment to the defense of a vaguely defined 1,000-mile SLOC zone off Japan, South Korea also began to explore the importance of SLOCs to the Republic of Korea's economic security. The Korean air force and navy have increased their surveillance of Soviet activity in the

Eastern Sea (Japan Sea).[66] Numerous South Korean officials and scholars have echoed American concerns about the growth of the Soviet Pacific Fleet headquarters in nearby Vladivostok, Soviet access to Vietnamese bases, and possible improved access in North Korea. A small but significant organization emerged in Seoul called "The SLOC Study Group-Korea," affiliated with equivalent groups in the United States, Japan, Australia, and Taiwan.[67] South Korea has two major problems in addressing the SLOCs issue. One is the bureaucracy's response to the security of South Korea. Because the overwhelming threat emanates from North Korea via land and air, it has been almost impossible to deny the ROK Army and Air Force the lion's share of South Korean resources. As a result, South Korea does not possess major naval forces—but then, neither does North Korea.[68] Consequently, the positions of South Korea and Japan on SLOCs have become increasingly congruent. Both recognize the importance of far-flung sealanes as their economic lifelines, both are capable of creating the naval forces to fulfill expanded roles if they decide to, and both see SLOC defense in distant areas as a shared security interest about which neither can do much and both remain dependent on the United States. Both talk a better game than they can play. This may not seem like commonality, but compared to the issues over which they disagree it is a significant area of agreement upon which future cooperation may be built.

If Korea-United States-Japan cooperation is an area of transitory disagreement, the last of the four major problem areas is even more ephemeral though less obviously contentious. This is the triumvirate of arms control, disarmament, and tension reduction. On the surface and in principle, the United States and the Republic of Korea largely agree about the need for all three. No one wants to be against such meritorious objectives. However, the United States and South Korea approach these related issues from very different perspectives, derived from their respective roles as a superpower and a relatively small regional power, respectively. Following the Reykjavik and Moscow summits, the United States stressed to all its allies that they would not get short shrift in any superpower arms control and/or reduction deal that might emerge from United States–Soviet negotiations. Asia has received full attention from the United States.[69] However, the Soviet Union also has been paying attention to Asian nuclear concerns.

Gorbachev's July 1987 expanded proposals, to do in Asia what Moscow and Washington were then on the verge of doing in Europe, were deft diplomatic overtures.[70]

There is a fairytale quality to these offers and counteroffers when juxtaposed in their global and regional contexts. The nuclear balance between the United States and the Soviet Union, whether seen from a European or Asian viewpoint, does not focus in the Atlantic or the Pacific, but in the Arctic Ocean. The ultimate negotiations center on that realm, not on the NATO or East Asia theaters that so often are seen as primary. Nonetheless, because most people live in Europe and Asia, where theater nuclear disarmament looms large, these tend to be the burning issues. Thus, South Koreans—like Japanese and Chinese—are concerned that nuclear arms reduction in Europe might create conditions in which the Soviet Union could redeploy additional forces to Asia. The continued threat of that shift fosters circumstances in which the superpowers are seen as chess players and regional states as pawns. Whatever the ultimate outcome of the nuclear arms reduction/control talks, this is the message retained among South Koreans. Seoul has little or no say in the immediate area concerning its own fate. Tokyo and Beijing have some voice, but Seoul does not; and Pyongyang does not even register as an actor. Consequently, both global and regional nuclear disarmament negotiations make Koreans feel much like interested bystanders—not really consulted, but advised after the fact. Koreans, South and North, legitimately feel like proverbial shrimps among whales.

The Prospects for Continued Cooperation

Korea-focused disarmament and tension reduction is properly the province of Koreans. However, here, too, there is an air of unreality. Both Koreas engage in ostensibly serious talks aimed at tension reduction and unification. They may someday succeed, if North Korea's obdurateness can be overcome. It is clear that Pyongyang's policies are the major obstacle, but the prospects are not helped by differences in United States and South Korean attitudes toward tension reduction. Issues in North-South negotiations are backed by the major powers in ways that make clear their confidence that the two Koreas cannot resolve their differences.

Consequently, the United States and South Korea, while apparently backing roughly the same proposals, actually hold to radically different perspectives. South Korea (like North Korea) claims to be sincerely pursuing tension reduction and unification but does not always act accordingly. Behind the facade of United States support for the Republic of Korea's formulas, most Americans working on Korean affairs are openly cynical about the feasibility of tension reduction. Despite that pessimism, the United States periodically makes serious moves toward tension reduction. So far, they have been of little avail in terms of reducing North-South stresses, but they have agitated officials in Seoul. Washington's efforts at "smile diplomacy" in 1983 and 1987 were examples of that. Fortunately, in mid-1988 President Roh Tae Woo announced a policy shift that gave a green light to improved United States–North Korean relations. This could signal a significant change in United States relations with both Korean states.

Like unification, prospects for real arms reduction and tension control in contemporary Korea are considered by most American and South Korean officials to be dim, at best. Nonetheless, because both the United States and the Republic of Korea are required for domestic purposes to keep these issues high on their respective policy agendas, they are compelled to cope with each other's perspectives and notions of how best to resolve the problem. This situation shows no sign of change.

It is worthwhile recalling the testimony of Assistant Secretary of Defense Richard Armitage briefly cited earlier. Armitage forecast in early 1987 that the United States and Korea were "about to embark on a new security partnership." He said optimistically, "In only a few decades [South Koreans] have transformed a nation that was completely dependent upon the US military presence into a self-confident, newly industrialized country that is capable of assuming most of the cost of its own defense."[71] Korean Defense Minister Yi Ki Baek confirmed that assessment at the Nineteenth SCM in May 1987.[72] In the warm glow of such harmony it is easy to become upbeat. For forty years prior to the United States' arrival in 1945, Korea was oppressed under Japanese rule. In the roughly forty years the United States has been its mentor, South Korea has become a more viable state geopolitically. It is easy to gloss over the security differences cited here and assume that harmony will prevail in the future. However, those assumptions must be evaluated against the trends toward the maturation of North and South Korea, the internationalization of the Republic of Korea, the broadening of North Korea's horizons, and the re-Asianization of both Koreas.

The context of United States–Korean security relations is changing rapidly. Though South Korean Defense Minister Oh Ja Bok in mid-1988 advocated keeping United States forces in Korea until the early twenty-first century,[73] that seems very unlikely—at least in their present configuration. Too many factors are changing too rapidly for there to be much prospect of retaining United States forces in Korea indefinitely. The global context in which United States–Korea security relations exist is rapidly changing. The economic, political, diplomatic, and strategic circumstances that define South Korea's role in the world are being altered by successes and innovation in ways that make United States forces there less useful and less desirable.

The key issue on the bilateral security agenda is to determine the criteria necessary for both countries to change their relationship. The United States' postwar programs in Korea have largely succeeded economically, politically, and strategically. American allies' (including Korea's) successes have produced capabilities to fulfill larger roles and responsibilities. To the extent they do not do so, they are behaving unfairly toward their American benefactor. This does not mean the United States is a victim of any sort of perfidy or manipulation by its allies, but that it may become a victim of its own naivete and gullibility in not pressing as assiduously as it might for greater fairness and justice in collective security arrangements with increasingly prosperous allies. In this context, both the United States and the Republic of Korea need to devise a gradual shift of defense roles to the Republic of Korea and away from the United States. Though long overlooked and largely forgotten, there is a "sunset clause" in the United States security commitment to Korea. The United States is gradually working itself out of a job in Korea's defense. Incrementally in coming years, the United States should wrap up loose ends and transfer the responsibility and cost of Korean security to our Korean allies.

This does not mean that the United States should suddenly abrogate its security commitments to South Korea—the cut-and-run scenario—but suggests that Washington should accelerate the transfer of burdens and costs to Seoul. Nor does it contradict the logic or desirability of increasing bilateral cooperation on regional security issues that are central to both countries' national interests. Some American roles within Korea may legitimately remain a while longer, but they need not be as large as they now are. It is time for the United States to begin invoking the sunset clause in our security ties with Korea and moving on. We should bear in mind, however, that horizons are being recast with each passing day as China's ambivalent reforms and the Soviet Union's more significant shifts alter the international environment. Both the United States and the Republic of Korea have to confront the ways in which their ties with China and the Soviet Union are rapidly changing. In particular, Seoul's dramatic shifts in foreign policy—guided by its "Nordpolitik" and post-Olympics flexibility—are injecting very different elements into the equations that determine the nature of the external threat, what burdens have to be borne, and who should bear them.

NOTES

1. The author examines broader facets of United States–Korean relations in his *US Policy and the Two Koreas* (San Francisco and Boulder, CO: World Affairs Council of Northern California and Westview Press, 1988). Small portions of this study are drawn from that analysis.
2. *Mutual Defense Treaty*, Treaties and Other International Acts Series 3097, U.S. Department of State publication 5720 (Washington, D.C.: Government Printing Office, no date).
3. See, for example, the opinion of Gregory Henderson, *Christian Science Monitor*, 5 October 1983, p. 18.
4. *Wall Street Journal*, 2 February 1981, p. 21.
5. *Washington Star*, 27 June 1981, p. 9, quoting Chun's extreme remarks at a news conference in Jakarta, Indonesia.
6. *Christian Science Monitor*, 25 April 1985, p. 3.
7. *Yonhap* wire service, 1 April 1986, Foreign Broadcast Information Service, East Asia and Pacific (hereafter FBIS-EAS) 4, 2 April 1986, p. E1.
8. *Korea Newsreview*, 2 August 1986, p. 8.
9. *Christian Science Monitor*, 11 December 1986, p. 24.
10. The writer examined the reasons not to assume the worst in "Keeping North Korea out of Soviet Hands," *Far Eastern Economic Review* (hereafter FEER), 14 May 1987, pp. 40–41.
11. Not all South Koreans or Americans fit the rough picture suggested here. One outstanding example of a South Korean defense analyst who is far less parochial than many of his colleagues and is empathetic with broader United States perspectives is Dr. Cha Young Koo. For examples of his analyses, see "Strategic Environment of Northeast Asia" in *Korea & World Affairs*, Summer 1986, pp. 278–301, and "North Korea's Strategic Relations: Pyongyang's Security Cooperation with Beijing and Moscow," in Robert A. Scalapino and Lee Hong Koo, eds., *North Korea in a Regional and Global Context* (Berkeley: University of California, Institute of East Asian Studies, 1986), pp. 371–86.
12. FBIS-EAS 4, 11 October 1979, p. E1.
13. *Pacific Stars & Stripes*, 23 August 1986, p. 13.
14. *Korea Herald*, 10 May 1987, p. 1, and the Monterey, California, *Herald*, 7 August 1988, p. 2A.

15. *Korea Herald*, 11 March 1987, p. 3.
16. For coverage of their respective reassurances, see *Christian Science Monitor*, 8 February 1983, p. 3; *Washington Post*, 30 March 1982, p. 8; *Chosun Ilbo*, 13 May 1984, p. 2; *Christian Science Monitor*, 1 April 1986, p. 5; and *Korea Herald*, 28 February 1987, p. 3.
17. See *Washington Post*, 4 February 1981, p. 18; and *Wall Street Journal*, 5 February 1981, p. 24.
18. *Korea Herald*, 4 April 1987, p. 3.
19. Ibid., 26 May 1984, p. 1.
20. *Korea Newsreview*, 22 November 1986, p. 7.
21. *Korea Herald*, 13 February 1987, p. 3.
22. Ibid., 8 May 1987, p. 1, and 10 May 1987, p. 3.
23. Ibid., 17 May 1987, p. 1.
24. *Korea Newsreview*, 9 August 1986, p. 6. The author was routinely queried on the "rent" issue during a July 1988 lecture tour of Korea for USIA.
25. *Korea Newsreview*, 17 January 1987, p. 14.
26. The author strongly reinforces that message in his *US Policy and the Two Koreas*.
27. *Korea Herald*, 2 April 1987, p. 3. See also the work being done by the "Sea-Lane Study Group," headed by Dr. Kim Dal Choong.
28. Defense Minister Yoon Song Min answered opposition party charges that United States restrictions on technology transfers were crimping Republic of Korea arms sales by threatening to diversify weapons imports, thereby reducing the United States' ability to inhibit South Korean relations with third states. *Korea Herald*, 28 November 1985, pp. 1–2.
29. Colonel/Dr. Hwang Dong Joon, Korea Institute for Defense Analysis and Visiting Fellow, Asian Studies Center of the Heritage Foundation. "South Korea's Defense Industry: An Asset for the US," Heritage Foundation Backgrounder no. 38, 10 December 1985.
30. FBIS-EAS 4, 4 April 1986, p. E1.
31. FEER, 26 February 1987, pp. 32–33.
32. *Time*, 6 August 1984, p. 19; and *Korea Herald*, 27 July 1984, p. 3.
33. FBIS-EAS 4, 7 April 1986, p. C2, and 30 July 1986, p. E1.
34. FEER, 19 March 1987, pp. 14–15. For Pyongyang's early com-

ments, see *Nodong Shinmun*'s commentary in FBIS-EAS 4, 25 July 1985, pp. D2–3.

35. *Christian Science Monitor,* 18 November 1986, pp. 1 and 56, and 19 November 1986, p. 9; and *Time,* 1 December 1986, pp. 34–35.
36. *Korea Herald,* 1 March 1987, pp. 1–2, and 3 March 1987, p. 5.
37. Brigadier General John C. "Doc" Bahnsen, Jr., USA, *Armed Forces Journal International,* November 1985, pp. 78–88.
38. *Korea Newsreview,* 13 June 1987, p. 6.
39. *Washington Post,* 2 May 1983, p. C13.
40. *Korea Herald,* 6 June 1984, p. 1.
41. *Bulletin of the Atomic Scientists,* April 1985, p. 4.
42. FBIS-EAS, 22 April 1985, p. D12, and *Korea Herald,* 20 April 1985, p. 1.
43. For examples of these allegations, see FBIS-EAS 4, 19 July 1984, pp. D1–5, and 3 April 1987, p. D4.
44. *Korea Herald,* 12 October 1984, p. 1.
45. *Christian Science Monitor,* 3 September 1986, p. 13.
46. See, for example, *Tong-A Ilbo,* 3 May 1983, p. 3, and *Chosun Ilbo,* 25 January 1983, p. 2, and 4 May 1983, p. 2.
47. See the *Chungang Ilbo*'s editorial, 9 June 1984, p. 2.
48. One officer's delineation of the problem and proffered solutions are in Brigadier General Rhee Taek Hyun (ROKA), *US-ROK Combined Operations* (Washington, D.C.: National Defense University Monograph Series, 1986).
49. See, for example, Gregory Henderson, "The Institutional Distortion in American-Korean Relations," *Korea Scope,* June 1982, pp. 3–17.
50. For one reporter and U.S. embassy staffer's experience with the Republic of Korea military's attitudes, see FEER, 20 June 1985, p. 28.
51. FEER, 12 March 1987, pp. 22–25. For a useful summary of the official perspectives, see ROK Army Colonel Lee Suk Bok, *The Impact of US Forces in Korea* (Washington, D.C.: National Defense University Press, 1987).
52. The author has suggested approaches to redesigning the CFC in his *US Policy and the Two Koreas,* and "Korean Politics and US Policy: Higher Pressure and Lower Profile," *Asian Survey,* August 1987. See, also, Gregory Henderson, "Time to Change

the US-South Korea Military Relationship," FEER, 24 September 1987, pp. 36–38.

53. For an excellent Japanese assessment of these problems, see Kataoka Tetsuya, *Waiting for a Pearl Harbor* (Stanford: Hoover Institute Press, 1980), and "Japan's Defense Non-Buildup: What Went Wrong?," *International Journal of World Peace*, April–June 1985, pp. 10–29. For a more cautious Japanese assessment, See Okazaki Hisahiko, *A Grand Strategy for Japanese Defense* (Lanham, MD: University Press of America and Abt Books, 1986). The author addressed these troubled defense and economic relations in *US-Japan Strategic Reciprocity* (Stanford: Hoover Institution Press, 1985).

54. FEER, 6 August 1987, pp. 12–13.

55. The author addresses the reasons why, and their impact on the United States, in "US-Japan Security Relations: The Case for a Strategic Fairness Doctrine," in Ted Galen Carpenter, ed., *Collective Defense or Strategic Independence? Alternative Strategies for the Future* (Washington, D.C., and Lexington, MA: Cato Institute and Lexington Books, 1989).

56. The quote is from Jean-Pierre Bechter, secretary of the French parliament's National Defense Committee, in *Washington Post* (National Weekly Edition), 27 July 1987, p. 17.

57. FBIS-EAS 4, 11 February 1982, p. C1.

58. *Korea Herald*, 6 July 1986, p. 4.

59. Ibid., 13 June 1985, p. 1.

60. There have been many. See, for example, ibid., 26 November 1983, p. 8, and FBIS-EAS 4, 9 December 1985, p. C4.

61. For early examples of this sort of advocacy, see *Hangook Ilbo*, 10 June 1982, p. 2, and 25 September 1982, p. 2; *Kyonghyang Shinmun*,8 September 1983, p. 2, and *Korea Herald*, 6 September 1983, p. 1.

62. *Korea Herald*, 7 February 1985, p. 4.

63. For two of the many examples of North Korean attacks on this notion and on joint efforts to blockade the straits between Japan and South Korea, see *Nodong Shinmun* commentaries in FBIS 4, 8 August 1986, p. D7, and 17 November 1986, pp. D3–4.

64. FBIS-EAS 4, 9 November 1983, p. E6, and 30 October 1985, p. E1, and *Korea Herald*, 31 October 1985, p. 1.

65. *Korea Herald,* 19 September 1984, pp. 1, 7; and FBIS 4, 17 September 1984, p. E8.
66. *Korea Herald,* 13 May 1987, p. 1.
67. *The SLOC Study Group—1987,* pamphlet (Seoul: Yonsei University Institute of East and West Studies), and its program for the Fifth International SLOC Conference, "Resources, Maritime Transport, and SLOC Security in the Asia-Pacific Region," 15–17 June 1987, Seoul. Also see *Korea Herald,* 18 June 1987, p. 3.
68. U.S. Naval Institute *Proceedings,* March 1987, p. 67.
69. See Edward L. Rowny, *Arms Control: The East Asian and Pacific Focus,* Current Policy Series no. 904 (Washington, D.C.: U. S. Department of State, January 1987).
70. FEER, 6 August 1987, pp. 10–11.
71. *Korea Herald,* 28 February 1987, p. 3.
72. FEER, 21 May 1987, pp. 34–35.
73. FEER, 23 June 1988, pp. 11–12. Surprisingly, a poll indicated a 73 percent majority of South Korean college students supported keeping United States forces in Korea, even as they found many objectionable aspects to their presence. See *Korea Times,* 27 July 1988, p. 3.

FIVE

Taiwan: Some Problems of Foreign Relations and Security

Harry G. Gelber

During the middle months of 1989 it became evident that China had entered a period of transition that may well turn out to be another of the sudden and major turning points that have marked China's history during the last couple of centuries. Under these circumstances, many if not most of the elements in an assessment of the political and security positions of not only the People's Republic of China but also the Republic of China must be accompanied by question marks.[1] The relationship between these two, in "one country, two systems," is one of increasing subtlety and complexity; and these complexities seem certain to grow following the appearance of what can only be called a popular revolt on the mainland and the moves to suppress it. Whatever the immediate outcome of these events, the authority of the ruling elites in Beijing has been compromised, perhaps permanently. In consequence, much of the framework for any consideration of Chinese affairs in the final decade of the twentieth century may need to be reconsidered.

Republic of China's Economic Development

Taiwan's economic development has to be seen against the background of a gradual if uneven drift towards democracy that seems to have accelerated in recent years, for instance with the formation of the opposition Democratic Progressive party following the lifting of martial law in July 1987. Fundamentally, as Yu-shan Wu has pointed out, the Kuomintang has been trying to transform itself from "an ideology-based revolutionary party into a political machine geared to electoral competition" and has sought "a grand shift of major legitimacy principle from traditional reciprocity to a modern political contract."[2] All that has also involved a transition in domestic politics from the Chiang dynasty to a civilian Taiwanese president with an increasingly Taiwanese bureaucracy, which has gone not just well but much more smoothly than might have been hoped. After the Republic of China's elections of December 1989, roughly half of the legislature will have been elected in Taiwan. No less satisfactory has been the reform of the bureaucratic structure and the infusion of new blood.[3] In addition, the economic base on which the authorities in Taipei can now operate is one of great strength and considerable versatility.[4] Growth has been continuous from the early 1950s and has, more recently, achieved remarkable levels. That has been due to a variety of factors, including United States aid; Taiwan's links with other economies, and especially the American one; the adherence to export-led growth; and, above all, the domestic social and economic environment. This has included a dedication to hard work, a reservoir of cheap labor, enthusiasm for education and training—as one factor in a stress on constant improvement in factor productivity—and various structural changes. Large numbers of young people have gone abroad to get high-level professional or technical training, often in the United States, as part of the government's general stress on technological progress and productivity. Underlying all that has been the agricultural land reform of earlier days, high rates of savings and investment, anti-inflationary disciplines, and, not least, some wise planning. As one observer has put it,

> In the postwar period, Taiwan has been fortunate to have had the service of a small group of extraordinarily able, experienced and dedicated technocrats who helped to shape and implement its highly successful development strategy.[5]

By the later 1980s the Taiwan authorities were encountering major difficulties created by the country's successes.[6] The Republic of China's foreign currency reserves had reached a level of some $75 billion, perhaps the second highest level in the world. The Taiwan dollar rose markedly—some 45 percent in two or three years—threatening inflation, while at the same time traditional trading partners began to complain loudly about unequal trade and persistent trade imbalances. Taiwan's businessmen therefore embarked on a diversification of trade everywhere, including with the socialist bloc; foreign investment programs in a variety of places, including the Philippines and Thailand; and the moving of factories overseas. This program took advantage of relatively cheap overseas labor and the desirability of diversifying assets in a new program made attractively cheaper by the rise of the Taiwan dollar. At the same time the new program tended to head off some foreign complaints about the nation's trading policies and to strengthen Taiwan's entire economic and investment base. Amid these shifts the government has shown itself especially anxious to keep down inflation and price rises and, in addition and more recently, to address the large problems of the entire domestic infrastructure and social services.[7]

Price rises have been kept in check by tariff cuts, cheaper imports resulting from the rise in the Taiwan dollar, and a lowering by the government of fuel and energy prices. At the same time, the policy of protecting Taiwan's exporters by keeping the currency appreciation slow and steady has allowed currency speculators to make large and certain profits. Money has therefore flooded in, and the money supply rose sharply during 1986–89.[8] Much of the resulting pressure on prices went into real estate and the stock market. In 1988, the Taiwan equity index rose by almost 280 percent. On the other hand, the fact that Taiwan's economy had been so strongly export led[9] also tended to make it vulnerable to that same rise in the Taiwan dollar, as well as to the possibilities of any serious political changes or economic slowdown in major markets abroad, especially in Japan or the United States, which takes around 40 percent of Taiwan's exports.

What the government has only begun to address in a major way—including in this year's budget—are the problems of Taiwan's domestic infrastructure, which remains poor and therefore throws, among other things, additional strains on business and society at large. The difficulties include increases in the price of land, a shortage of industrial land, and severe pollution problems.[10] These things have created a need for industrial and urban reform in ways that are bound to be costly. Together with all that has come growing labor militancy and resentment by farmers, particularly over the rising price of land. In the latter part of 1988 the government announced an enhanced program to buy land for infrastructural projects. General government expenditures are being sharply increased, with some emphasis on deficit financing in spite of the threat of inflation. The budget brought down this year by the central government shows an almost 25 percent increase over 1987–88. Much of the expansion seems due to increased spending on social security and public investment for economic development, education, and science, which will absorb no less than 15 percent of the national budget.

In more general terms, the government clearly intends to move towards a more efficient—and, so far as possible, self-sufficient—system of energy use, a move from labor-intensive to hi-tech and science-based methods of production, and much greater attention to the domestic infrastructure, including much of the road, rail, port, and airfield network. The government will no doubt continue along such lines while trying to keep inflation in check and attending to major popular concerns about social matters.

Security Concerns

Prominent among the many reasons for Taiwan's drive for economic growth has been the recognition that growth is a major precondition for military strength. The security concerns of the Republic of China come in a considerable variety of shapes and sizes. The principal, indeed overwhelming, concern is that vis-à-vis the People's Republic. The Chinese civil war has never been officially ended. The single most important matter on which Chinese on both sides of the Taiwan Straits are in almost full agreement is that there is only one China (although many Taiwanese strongly disagree). The only dispute is who, and which system, should prevail or govern within it. Hence, "one country, two systems." Beijing has repeatedly expressed a preference for a peaceful resolution of differences but has not finally renounced the use of force, partly in case that should diminish the incentives for the Republic of China to negotiate on the subject of reunification. Indeed, such a renunciation might even, in the view of Beijing, remove threatened sanctions in the event of a unilateral declaration of independence by the authorities on Taiwan. Beijing has given five grounds on which it would consider the use of force against the Republic of China justified: in the event that the Republic of China acquired nuclear weapons; in the event that it attempted a unilateral declaration of independence; if the Republic of China sought a strategic alignment with the Soviet Union; if there were serious political instability on Taiwan; or if there were a "protracted refusal to negotiate" with the People's Republic on reunification. For the moment, however, Beijing has stopped the symbolic bombardment of the offshore islands and withdrawn forces from the Taiwan Straits. The People's Liberation Army (PLA) is not currently deployed in such a way as to threaten Taiwan, or even to suggest an intention to threaten. Yet it remains that the People's Republic has nuclear forces, a substantial surface and sub-surface navy, and a large if somewhat antiquated air force.[11] Moreover, the People's Republic has enjoyed a decade of economic growth that, of necessity, has strengthened the sinews of its military power, while the PLA has moved towards modernization and reform in a way that must also give the military on Taiwan cause to think.[12] Whether these trends will continue following the political upheavals of 1989 in Beijing and the changes in Beijing's relationship to the global balance of power is, however, a very different question.

The chief Taipei concern seems to remain that of an invasion by the People's Republic, not because it currently seems particularly likely, or even feasible. The People's Republic has an amphibious capability that could at most lift some 10,000 men.[13] That, plus the general navy and merchant navy backup available to Beijing, would appear to be a totally inadequate basis for an attempt at invasion. Nevertheless, this constitutes the ultimate threat to the Republic of China's independent survival. And no doubt the further development of People's Republic forces, including the maintenance and exercising of power projection capabilities that include some seaborne landing forces, has been duly noted in Taipei. Nor does air or airborne attack by itself seem very likely, given the persistent qualitative advantages enjoyed by Taiwan's air forces in spite of their numerical inferiority.

Some kinds of naval action against the Republic of China's lines of communication might be more plausible. These could include attempts at blockade, either directly by surface interception of shipping or by mining the approaches to Taiwan's chief harbors.[14]

Interruption of shipping links, if not by the sinking of many vessels, then by the threat of mines, could cause severe economic difficulties on Taiwan almost immediately. To achieve that, it would be necessary for the People's Republic to to cut off Taiwan's communications entirely, or to rely on energy starvation on the island (after all, some 50 percent of the Republic of China's energy is supplied by nuclear power). Even if few ships were sunk, there would be dramatic increases in shipping insurance rates, a sharp decline in the willingness of many ship-owners to let their vessels enter Taiwanese waters, and severe damage to Taiwan's trade. Of course Beijing would incur heavy costs by using such tactics. Not only would the United States and the West be alienated, with severe consequences for Chinese trade and technology acquisition policies, but all of Asia would be affected by the revival of ideas about "Chinese expansionism." Nevertheless, the potential for blockade remains. It is therefore interesting to observe that the Republic of China has a mere eight minesweeping vessels and few ancillary capabilities.[15] If that condition is intended to convey the impression that such matters can safely be left to the United States Navy, that could turn out to be a high-risk policy in some circumstances.

But the Republic of China's security concerns are much broader than those relating to any short-term danger of attack from the mainland, or even from a clash between the People's Republic and the Republic of China per se. For example, Taiwan must watch with care not just the political implications of the slowly improving Sino-Soviet relationship[16] but also its military consequences. Insofar as that improvement lowers the demand on the Chinese side for large deployments on China's northern borders, it also eases the difficulties hampering the reform and upgrading of the PLA. Or again, the Republic of China has made increasing gains in trade and economic exchanges with the People's Republic. Although economic relations remain nonexistent at the official level, a very great deal has been done by way of nongovernmental exchanges, for example, through Hong Kong. I return to the point below. For the moment it seems worth stressing that from a security point of view, these unofficial relations give Beijing a stake in Taiwan's prosperity. At the same time it would be clearly unwise to allow any major sector of Taiwan's economic activities to become unduly dependent upon relations with the People's Republic, lest Beijing should acquire significant economic or political leverage over Taiwan.

For the same reason, security issues would be involved in any erosion of Taiwan's political relations with others and especially with the United States. Then, too, the Republic of China will probably be wary of any further expansion of Japan's military capabilities. From Taiwan's point of view, Japanese military power is doubtless helpful in the sense of establishing or maintaining a balance in the seas and air space between China and Japan. But there would surely be concern about any capabilities, or any signs of intentions, which could suggest the revival in Japan of the attitudes and views of the 1930s. More important, no doubt, is the wariness about the size, efficiency, and role of the Soviet Pacific fleet and, more particularly, the balance between it and the U.S. Navy. That issue feeds into the more general one of Taiwan's dependence upon free passage over the oceans both near to and far from the island. Any threats to that freedom of passage, whether from the Soviet navy, through a command by Japan of the seas adjacent to Taiwan, or by the threat of interruption by some third power, would at once cause great alarm. All that is, of course, in addition to the danger of any new war, for example, in Korea, which could involve substantial Republic of China interests.

The other side of that type of concern is the way in which declining dangers of conflict might erode the help and support, especially of the United States, to which Taiwan is accustomed and on which it relies. It is, after all, clear that in terms of United States policy the People's Republic of China has over the years scored a number of victories over the Republic of China, that Beijing would prefer that the United States pressure Taipei to negotiate more intensively over reunification, and that it would like to see the Taiwan Relations Act repealed. And against that, it remains true that Taiwan has very considerable strategic importance, not solely for her own citizens but also for the major powers involved in Northeast Asia.[17] Taiwan lies astride the Taiwan Straits, commands the southern access routes to Japan, and dominates the Eastern Sea approaches to the People's Republic. It is an important link between the Pacific and Indian Oceans. Its loss would break the line of islands, all friendly to the West, that lie off the East Asian mainland and that have since the days of Dean Acheson been regarded as of the greatest importance for the West's strategic posture in the Pacific region and beyond. A secure Taiwan is, from a United States point of view, of special importance in a sea area where over the last decade or more the strength of the Soviet navy has very greatly increased.[18] And, perhaps not least, the Republic of China has been a reliable ally of the United States for almost forty years, a record not to be cast lightly aside by either party.[19] Nevertheless, the slow improvement of Sino-Soviet relations and the more constructive attitudes in both South and North Korea mean that the threat environment in the areas surrounding Taiwan can be said to have become much more benign. That, together with United States policy developments in other parts of the globe and, indeed, shifts in American domestic opinion, seems likely to lead to some reduction in United States forces levels deployed around the Western Pacific. If so, Taipei must be concerned about the force of its continuing appeals to Washington for defense and other assistance.[20] The Republic of China has for long based its claims for Western support on the idea that it was the "free world's 'unsinkable aircraft carrier' in the Western Pacific."[21] But in the new circumstances it may be that this function is no longer so urgently needed.

In the light of all these factors, it is natural that Taiwan should be looking to develop its defense capabilities further. However, this thrust of security policies must be seen in the context of broader areas of policy, especially foreign relations and economic, trade, and technology policies. The Republic of China has had considerable success in using economic diplomacy to make friends and influence people without directly challenging the embargo that the People's Republic has tried to place on the ability of others to have formal diplomatic relations with Taipei. Taiwan's foreign investment has clear implications for the interest that other countries will take in the security of the Republic of China. Taipei is establishing preliminary links with the OECD, has established an interbank loan market, and made clear its intention to try to match Hong Kong as an international financial center.[22] The policy of technological upgrading and the establishment of partnerships of various kinds with foreign high-technology entities also have implications for both the Republic of China's international clout and the technical enhancement of its armed forces.[23]

The Republic of China already has military forces that are relatively well trained, well equipped, and have high morale. The fundamental defense requirements are not difficult to discern. As an island state, Taiwan must try to have an army capable of repelling invasion and a navy and air force capable of keeping hostile sea and air forces at a distance. The requirements that flow from that—and leaving aside questions of alliance with foreign powers—include a powerful army, an air force strong in fighter and fighter-ground attack aircraft, shorter-range air-to-air and air-to-surface missile capabilities as well as surface-to-air missiles for home defense, and a navy able to achieve at least sea denial along the approaches to Taiwan. Together with that should go a policy of steady technical upgrading.[24] At present Taiwan's order of battle includes active armed forces of 405,500, backed by reserves of over 1.6 million.[25] The active forces include 270,000 army personnel organized into eighteen light and heavy infantry divisions and two motorized ones. Their equipment includes over 1,200 tanks, 300 of them medium tanks, and 1,600 surface-to-air missiles. The navy has 26 destroyers and 10 frigates as well as over 50 missile patrol craft. The air force has some 300 F-5 and F-104 fighters, many of them equipped with Sidewinder air-to-air missiles. Republic of China forces could therefore expect to present significant challenges for any attacker.[26]

These capabilities are in a process of constant development, which includes policies to increase self-reliance where possible. After the United States decided in 1982 to reduce the dollar amount of its military sales to the Republic of China and not to sell high-technology weaponry at all, Taiwan was forced to pursue policies of improved self-sufficiency. That has had, and will have, multiple benefits. The pressures for improved military hardware can be expected to stimulate industrial technologies as, for instance, in electronics. A strong domestic defense industry complex reduces the possibilities of certain foreign pressures on the Republic of China and can improve exports through arms sales. Some of these benefits have already begun to appear.

As an island state, Taiwan must be especially sensitive to naval developments in the People's Republic. China's claims in the South China Sea, the need to protect its coasts in an era of growing naval capabilities around the Pacific, and the naval increases of its past and potential competitor, India, have all made it necessary for Beijing to pay special attention to its navy. That, in turn, must influence planners in Taipei. So in 1988 the Republic of China commissioned two Zwaardvis class submarines, bought from the Netherlands and renamed the Hai Lung class,[27] and has made clear its intention of building more such boats. Similarly, the China Shipbuilding Corporation has arranged to begin in 1990 the construction of eight Perry class guided-missile frigates. The U.S. Navy is to provide the blueprints and technical transfers for the first two, and all eight are scheduled to be in service by 1997.[28] For its air force, Taiwan has developed the AT-3, a training aircraft but one that might be able to carry bombs, and in 1988 Taipei introduced the prototype of its own, home-produced Indigenous Defense Fighter, which began a testing program in the following year. The plane, initially built with some help from General Dynamics, is planned to be capable of air-to-air and air-to-ground action, and Taipei is said to be planning to build 250 of them.[29] The Republic of China has also developed more advanced missiles. It now has some 54 naval patrol craft equipped with Hsiung Feng II missiles with greater range and speed than former ones, and in 1988 and 1989 Taiwan developed two new surface-to-air and air-to-air missiles.

In general, however, like the People's Republic,[30] the Republic of China looks to a policy on both weapons and training that puts stress on high quality rather than numbers of men on the ground. And great emphasis has been and continues to be placed on an intelligence collection effort that includes significant early warning capabilities.

The Republic of China and the People's Republic

There seems to be little doubt that at least until the events of May–June 1989 authorities in the People's Republic have been prepared to be flexible in practice about activities of the Republic of China. For instance, in international economics,[31] Beijing has been willing to look the other way on a number of relatively minor issues. Still, the feeling that Taiwan is necessarily and properly a province of one "Great China" would probably take precedence over most merely material interests if it were directly challenged. Beijing has gone to some lengths over the years to woo the people or the authorities on Taiwan. After 1978, with the transfer of formal American recognition from Taipei to Beijing, the People's Republic played down the antagonism between the two sides of China.[32] There were promises that the wishes and interests of the people on Taiwan would be respected and protected in any talks on reunification.[33] Taiwan would be able to retain its own political status quo, its ties with foreign investors, and even its own armed forces.[34] Deng Xiaoping himself indicated that it might be possible to combine a socialist system in the People's Republic with capitalist systems in Hong Kong and Taiwan even after reunification.[35] He suggested that no mainland military or administrative personnel would be stationed on Taiwan.

For the social and political structure of Taiwan, this created profound dilemmas. One of the fundamental emotions on both sides of the Taiwan Straits, as indicated earlier, is the sense of fraternity as members of a single Chinese civilization. It seems to be most marked among the older generation on Taiwan but is not absent among their juniors. For example, senior members of the Taiwan military look on the PLA with feelings that clearly include nostalgia, a sense of pride, and a tendency to applaud Chinese military achievements and successes, whether against the Soviet Union or Vietnam or India. As against that, history provides few reasons for the Communist party and Kuomintang to trust one another (and the events of May–June 1989 in Beijing are certain to increase such distrust on Taiwan). Nor will those outside Taiwan easily forget Beijing's crackdown in Tibet since October 1987 in response to Tibetan nationalism, or quell the increasing doubts about the fate of Hong Kong after 1997. Nor can they avoid worries about what might happen following some change in the leadership or political line in Beijing. Moreover, some 85 percent of the people on Taiwan are not Chinese but Taiwanese, and few of them are enamored of the idea of "reunification." Nor do these Taiwanese ignore the possibility that their interests might be overlooked in any negotiations between mainland Chinese in Taipei and Beijing. There is also some sense of separatism among many younger mainland Chinese who have spent their lives on Taiwan, have in many cases intermarried with Taiwanese, and see their future as lying on Taiwan. More specifically, there is strong anti-reunification sentiment directed at any suggestion that "one country" should mean submerging the prosperous Republic of China economy in the much poorer and less efficient economy of mainland China.

It is therefore not surprising that Taiwan authorities have been cautious in the extreme about negotiations and official contacts.[36] Where Deng promised local autonomy for Taiwan, Taipei responded by saying that, while reunification was desirable, it would only be possible if the People's Republic abandoned the communist system and adopted Sun Yat-sen's Three Principles of the People. President Lee Teng-hui has reiterated these principles[37] even while senior officials in Taipei speak of "mainland compatriots" and are careful, at various levels, not to affront Beijing's political sensitivities too directly.

On some issues, of course, views from Taipei and Beijing overlap in practice.[38] For both it is desirable that the Republic of China should flourish economically and technologically. For both it is desirable that economic and technical links between them be multiplied and strengthened. Although there are no official political links, the authorities in Taipei sent Finance Minister Shirley Kuo to attend a meeting of the Asian Development Bank in Beijing in May 1989, maintaining the fiction that this had no political implications.[39] While the government in Taipei maintains its official prohibition against Taiwanese investment in the People's Republic and forbids monetary and insurance institutions from doing business there,[40] the Republic of China is opening up to the mainland in other ways. Although the military threat to Taiwan has not been entirely removed, there is a very significant flow of visitors from the mainland, including family visits, across the Taiwan Straits.[41] Investors are, in practice, following suit. The estimates of Taiwan visitors to the People's Republic in 1988 range from 300,000 to 450,000. Among them have been scientists and scholars and possibly a hundred journalists. Taiwan consulting firms have made themselves responsible for group business visits. It is said that there may be up to a hundred Taiwan-owned shoe factories in Fujian Province alone, attracted by the low wages there; and there has been talk in Beijing of the possibility of setting up a special economic zone to cater to Taiwan's needs. After all, businessmen from Taiwan have a variety of ethnic, cultural, and linguistic advantages in dealing with the People's Republic.[42] At the same time, some categories of mainland visitors are allowed into Taiwan; and there is a small but steady stream of refugees paddling ashore to look for jobs and a better life after traversing the straits. There also seems to be some cooperation—not necessarily at official levels—

in making available to People's Republic citizens and groups the insights and especially the technological knowledge gained by Taiwan students abroad, especially those who have studied in the United States.

Although Beijing could hardly say so, the interests of the People's Republic and the Republic of China are far from being implacably opposed to each other, even in the matter of the American presence. Though Beijing continues to oppose the already diminishing flow of United States arms and supplies to Taiwan, it is in Beijing's interests also that Taiwan should feel secure. Insecurity would damage economic prosperity and the technology acquisition process from which the mainland also benefits. Moreover, if the Republic of China felt seriously insecure or likely to become markedly vulnerable to a People's Republic attack, other questions would arise. They could include a search by Taipei for other allies and possibly, ultimately, access by foreign troops to the island of Taiwan. In an extreme case it might even, given the technical and other capabilities of Taiwan, push the Republic of China into acquiring nuclear weapons. It is obviously in the interests of the People's Republic that nothing of the kind should happen.

In the meantime, the Republic of China is certain to watch very carefully the continuing expansion of the People's Republic's navy and especially of its blue water component.[43] The Soviet Union appears already to have given Beijing to understand that "Southeast Asia is in your sphere of influence as far as we are concerned" and this, if implemented,[44] is bound to have an impact on Taiwan's strategic outlook, including its assessment of the naval balance in adjacent waters. If the PLA does develop an aircraft carrier capability—and there has been some talk of that—then even if its primary purpose lies, as it surely does, in the South China Sea, it will have a significant impact on Taipei's defense planning.[45]

At the same time the whole question of relations—or non-relations—between the People's Republic and the Republic of China could well be profoundly affected by the "Beijing Spring" of May 1989 and its aftermath of repression in June and the months that followed. The first official reaction of the government in Taipei to the stirrings of democracy was to proclaim formally "to our mainland compatriots that we vow to act as the rearguard in their movement for freedom and democracy, and that we will join hands with them."[46] But then came the events in Beijing of June–July 1989. These were, among other things, an alarmed reaction against the logic of China's own reform program, against modernity itself. The leadership failed, in the end, to understand—or at least to accept—the organic links that must exist in the contemporary world between economic and technical developments and the needs of political liberalization. China's need for economic restructuring and technical advance depends upon, and must in turn strengthen, demands for freedom of ideas in other and adjacent areas. Education, some freedom of thought and expression, the encouragement of enterprise, cannot be promoted within the economic sphere and prohibited elsewhere.

Modernization was therefore always likely to place question marks against the present structure of the Beijing government, even the position of the Chinese Communist party (CCP) itself. Although the Beijing students carefully refrained from openly confronting the party, such a challenge was inherent in their movement. It was implicit in the spirit of the Beijing Spring, its language, its demands for something called "democracy," the symbolism of the occupation of Tiananmen Square, the obstruction of government business. It was also inherent in the students' demands, which, precisely because they were so general and indistinct, were therefore in principle unlimited. Never before had the government been so blatantly and pointedly defied, and in the heart of Beijing. It is, even in retrospect, not easy to see how the movement could have been tolerated, or its demands met, without undermining not just the authority of the government but the very basis of the party's claim to rule.

From the point of view of Deng and his colleagues, all this created horrendous dangers, not least of domestic disunity and therefore also of external weakness of just the kind from which they—and Mao—claimed to have rescued China some forty years earlier.

Whether that ruling group has fully understood, let alone discounted, the costs of repression may be another matter. The party and the PLA clearly remain divided on the crisis and its outcome,[47] even though debate, let alone open criticism, has died away. There does not now seem to be a legitimate process or institution for the transfer of power or even, in some respects, its exercise. Within the leading elements of the party, all that seems to remain is a fairly naked struggle for power, not always disguised by the surface appearance of solidarity and command. But between rulers and the ruled the old links have been broken, very likely for good. Nothing demonstrates more clearly the uneasiness of the leadership, behind the surface appearance of strength and confidence, than its alarmed and defensive reaction to the events of late 1989 in Eastern Europe, and especially in Romania. That reaction has been made obvious, inter alia, in the negotiations with Britain over the permissible degree of democracy in Hong Kong. The restoration of genuine political stability may have to await the succession to Deng, whose anointed successor appears to have no serious political base beyond Deng's own favor. Within the public at large, and especially in the major cities, disaffection has fairly clearly not evaporated just because it has been driven underground. There are signs of a significant loss of faith in the government and the Communist party; there is something like a crisis of legitimate authority.

No less important, in repressing the students and intellectuals the Beijing government has cowed into defensive silence precisely those classes and groups, not just of intellectuals but of entrepreneurs, researchers, reformers, innovators, not to mention managers and people more familiar with the outside world, who are desperately needed to fulfill China's economic and other aims. Those thousands of students who have gone abroad and found broader horizons are often reluctant to return to China and are likely to find frustration if they do. Some have gone to Taiwan. The number of students who will in future be allowed by Beijing to study abroad has been sharply cut.[48] There has been a marked revival of anti-intellectualism and anti-elitism, not least directed against economic reformers. The enrolling class at Beijing University has been cut from 2,000 to 800 entrants, 200 of whom are to be adults working in government agencies. All of them are to undergo a one-year ideological course controlled by the army before starting regular classes. Staff will also be given short intensive courses in Marxist thought. For all that Mr. Gorbachev and others, including many Japanese, have accommodated to the events in Beijing, economic and other contacts with the outside world are suffering and will suffer more. The decline in tourism may last for some time and will be especially damaging in cutting off hard currency supplies. The future of Hong Kong is subject to new doubts. Many international groups and firms have drawn down their commitments. There has been some switching of investment from Hong Kong to Southeast Asia.[49] Many loans to China have been deferred and higher interest rates are being charged.[50] Many high-level contacts and military sales have been cancelled.[51] China's financial and trading problems have increased, and there has been a clear, though just possibly temporary, turning inward of external and economic policies generally.

Taipei authorities have naturally taken the view that the repression in Beijing represents a real vindication of their long-standing views as to how China should be governed and what lines of social policy and economic development should be pursued.[52] President Lee denounced the events of June 4 as "this mad action" and remarked that the communist tyranny was "the shame of all the Chinese people of the world." While the Republic of China's government did not want to make itself vulnerable to any accusation that it had helped to organize the dissidents in Beijing, there was a good deal of spontaneous sympathy in Taiwan with the dissidents; and the government tried to help by passing information to the mainland and by offering refuge to students and others. But just what the consequences of the Beijing events will be for the Republic of China's security remains to be seen. An inward-looking and more ideologically oriented regime in Beijing might pay less attention to external relations but could also react more vehemently if its ideologies are contested. Damage to economic reform would not be good for China but would also slow down the growth of her military power. Economic cooperation with China via Hong Kong seems likely to become more difficult and less rewarding, but at the same time any temptations for Taiwan citizens to go too far in relations with the mainland will be diminished. On balance, one would guess that the physical and economic security of the Republic of China will not be greatly affected one way or the other.

The Republic of China and the United States

It remains to consider the relationship between the Republic of China and the United States, which has since the onset of the Korean War taken a practical interest in sustaining Taiwan's independence and welfare.[53] Taiwan obviously came to play a role of special importance to the United States during the cold war and the period of emphasis on "containment"—at first of the entire communist family of nations and, later, of China. The problems of containment became more acute during the Vietnam War with the help that Hanoi received from Beijing[54] as well as during the period of the Cultural Revolution, when the People's Republic seemed intent on containing itself by separation from the entire outside world. By the end of the 1960s things had changed dramatically. The increasing Soviet threat to China made a rapprochement between Beijing and Washington a necessity for the former and an opportunity for the latter.

However, the United States refused to abandon its long-time ally on Taiwan in this process. As President Reagan explained some years later,

> We don't believe that in order to make another friend, we should discard a long-term ally and friend—the people on Taiwan. I myself have said to some of the representatives of the People's Republic of China that we would think that they would have more confidence in us if they knew that we didn't discard one friend in order to make another.[55]

And the Nixon administration succeeded in establishing formal relations with Beijing by a formula that did not require cutting off of all ties with the Republic of China, even of official ones.[56] As matters progressed and the United States and China came to cooperate more closely in the containment of the Soviet Union and the fending off of any residual Soviet threats to China, the United States in 1978 gave the People's Republic full diplomatic recognition and ended the Mutual Defense Treaty with Taipei.[57]

Throughout these years Beijing attempted to increase the diplomatic isolation of the Republic of China while the authorities in Taipei mounted constant, low-level, and shrewd appeals to American public sentiment and to Congress.[58] In order to demonstrate continued concern for Taiwan and its welfare and to provide an appropriate legislative basis for continued cooperation, Congress passed the Taiwan Relations Act in April 1979, which in effect obliged the United States to supply defensive weapons to the Republic of China. Although its text is not precise as to the details of American concern for Taiwan's security, the legislation does make it clear that any effort to determine Taiwan's future by other than peaceful methods would be regarded by the United States as a threat to the peace and security of the Western Pacific and would cause very deep concern in Washington.[59]

Since then the People's Republic has repeatedly expressed its irritation with the flow of United States arms and equipment to Taiwan, and the flow has significantly diminished,[60] although the availability of American technology for the Republic of China's own arms manufacturers may have at least partly compensated for that. On 17 August 1982, a joint Washington-Beijing communiqué was signed, in which the United States agreed to reduce the quantity and quality of its arms sales to Taiwan gradually and eventually to cut them off. But Washington refused to give a definite cut-off date.[61] At the same time, Washington insisted that the gradual reduction in arms sales be made contingent on Beijing's promotion of peaceful reunification. That, the Americans said, would also obviate the need for more, and more advanced, weapons; and the combined effects of a reduction in weapons and a reduction of tensions would not, overall, curtail Taiwan's defense capabilities.[62] Of course the United States also continues to maintain commercial, cultural, and other links with Taiwan,[63] and a slow phasing out of United States arms supplies—as observed earlier—probably has the effect, among other things, of stimulating domestic production of many kinds of weapons and equipment in the Republic of China itself.

The United States clearly continues to have a number of important interests at stake in and with Taiwan. Washington must be interested in seeing that the Republic of China does not conclude other and potentially alternative alliances. For example, it would not necessarily be in the United States' interest (or, for that matter, in the Republic of China's) to have Taiwan become part of an explicitly Japanese sphere of influence. In particular, the United States would oppose any suggestion that foreign troops might come to the island. The United States will also want to see to it that the Republic of China does not suffer an economic downturn as a result of security problems. Nor should Taipei feel under the kinds of pressures that might cause it to develop, or otherwise acquire, its own nuclear weapons.

Per contra, it is very much in Washington's interests that the

People's Republic should not see Taiwan as a serious threat to mainland China and the source of a possible attack. More generally, it would not be in the American interest (or that of Japan) to have Taiwan become available as a naval facility or an aircraft carrier for the People's Republic. Indeed, if the facilities in the Philippines, at Clark and Subic Bay, were to become unavailable, and if, as a result, the remaining or substitute United States facilities in the Pacific region became geographically more scattered, it might be possible, subject to negotiation, to emplace some on the island of Taiwan. However, there is a major question mark here, which affects all notions of what larger security role the Republic of China might come to play in the Pacific region. On the one hand, the island clearly has a location of great strategic value, lying at the crossroads of many of the more important communication lines of the Western Pacific, of major potential importance to all its naval powers and to the control of its air space. But there are at least three major groups of difficulties that are likely to prevent the Republic of China from playing an important strategic role. The first is that any close military ties that Taiwan might develop with other Pacific powers would be certain to run into fierce opposition from Beijing, and in foreseeable circumstances few of them will be willing to incur Beijing's wrath in this way, the more so as such military links might be interpreted as a direct threat to China's coasts. Secondly, there is the likelihood that, in the increasingly volatile balance of power of the Pacific region of the 1990s, if the Republic of China were to form new and closer military links with one major power, that would mean incurring the potential or actual hostility of that power's competitors. Thirdly, precisely because Taiwan is of such strategic importance, it is probably in the interests of all major Pacific entities that it should be small, independent (de facto if not de jure), equipped defensively, and not able to act as a base for any of the major powers. That situation seems unlikely to change in the foreseeable future.

More narrowly, what if the United States terminated arms sales to Taiwan altogether? It is a prospect that Taipei must presumably take seriously in a period when the "ending of the cold war" is causing, and will continue to cause, a significant decline in America's willingness to accept external military or quasi-military commitments, especially ones that are relatively peripheral to the security perceptions of the wider United States public. Clearly such developments, together with the possibility of more assertive policies by Japan, could cause Taipei serious political difficulties. In the matter of military equipment, the government in Taipei would then have two chief options: to procure weapons from other sources or to develop their own defensive systems more rapidly. In recent years, the People's Republic has successfully frustrated Taiwan's attempts to purchase arms from sources like Brazil or the United Kingdom. Procurement from the Soviet Union may be a possibility.[64] As to the development of a domestic defense industry,[65] that is being pursued already but could be accelerated. After 1982 the Republic of China substantially increased its budget for military research and development, with tacit support from Washington.[66] In the new 1989–90 budget, defense will continue to get the lion's share of government spending, though as a percentage of government expenditure it goes down from 33.6 percent to 30.4 percent. Personnel will be cut by almost 9,000, and more emphasis is to be placed on modernizing weapons systems. This will include the procurement of new aircraft and naval vessels. No doubt these trends will continue, and the Republic of China will want to become as independent as possible in weapons production.

Conclusion

The two biggest clusters of questions relating to the Republic of China's political future and its security stem from the results of the 1989 repression in Beijing and the political repercussions on Soviet and United States policies of the 1989 East European revolutions. The answers to those questions have not yet emerged. But barring sudden fresh events of overriding and dramatic importance, one would guess that sudden or major changes are unlikely either in the relations between the Republic of China and the People's Republic of China or in those between Taipei and Washington. There may be slow and continuing movement in a period when Asian patterns of power will become more varied—even at times volatile—with a drawing down of United States interest in Taiwan, but without any abandonment of the United States insistence on peaceful solutions to the problems between Taipei and Beijing. No doubt that forthcoming American draw-down in the

Pacific will be viewed on Taiwan with some alarm unless or until it is shown that this does not really mean a large-scale American disengagement from the entire region. And though the sudden relative collapse of Soviet power could, other things being equal, give Beijing greater freedom of movement in the east and south, the precarious elements in the position of the Beijing leadership, and other domestic problems, seem likely to prevent rash forward movement by the People's Republic for the time being. And even a weakened regime in Beijing would be wise to let the relationships with Taiwan grow naturally at people-to-people and business levels, in ways that will be of fairly direct help both to ideas about "Great China" and to the economic refurbishing of the mainland. If and when the People's Republic does turn outwards again, when it eventually joins the ranks of the great powers, perhaps under new leaders, things will doubtless change. But that is unlikely to happen soon.

NOTES

1. For an earlier excellent brief review of China's posiiton on security questions, see Harry Harding, *China and Northeast Asia: The Political Dimension* (New York: University Press of America for the Asia Society, 1988).
2. "Marketization of Politics," *Asian Survey*, April 1989, pp. 382–400.
3. On state-society relations and the nature of Taiwanese society, see Thomas B. Gold, *State and Society in the Taiwan Miracle* (Armonk, NY: M. E. Sharpe, 1986). For discussion of earlier worries about Taiwan's internal affairs, see Trong R. Chai, "The Future of Taiwan," *Asian Survey*, December 1986, pp. 1309–23. For a more recent view, see Selig R. Harrison, "Taiwan after Chiang Ching-kuo," *Foreign Affairs*, Spring 1988, pp. 790–808.
4. For a general view of Taiwan's economic development, see Samuel P. S. Ho, "Economics, Economic Bureaucracy, and Taiwan's Economic Development," *Pacific Affairs*, Summer 1987, pp. 226–47. Also Chung-lih Wu, "Economic Development of the Republic of China: A Retrospect and Prospect and Studies," August 1987, pp. 72–100.
5. Ho, "Economics, Economic Bureaucracy, and Taiwan's Economic Development," p. 246.
6. For general recent reviews of the Taiwan economy, see *The Economist Survey of Taiwan*, 5 March 1988, and the *Far Eastern Economic Review*'s survey, "Awash in a Sea of Money," 5 September 1988, pp. 49–70. On financial and balance of payments matters, see Bela Balassa and John Williamson, *Adjusting to Success: Balance of Payments Policy in the East Asian NICs* (Washington, D.C.: Institute for International Economics, June 1987), pp. 57–67, and, at less professional levels, the Finance Section of the *Free China Review*, January 1989, pp. 4–29. Also useful are articles in the *Far Eastern Economic Review*, 15 December 1988, p. 127, and 29 December 1988, pp. 42–43.
7. On the need for increased public investment see, for example, Schive Chi, "An Economy in Transition," *Free China Review*, April 1989, pp. 4–9.
8. At one point in 1988–89 it was suggested that the money supply might be growing at a rate of 50 percent per annum.

On the other hand, the Republic of China enjoyed a very high rate of private savings.

9. The Republic of China's ratio of exports to GNP was estimated to be 48.5 percent in 1980 and 55.7 percent in 1988. *Australian*, 21 September 1989, p. 4.

10. For a discussion of the public resentment of pollution, see the *Economist*, 15 July 1989, pp. 63, 67.

11. This is of course not to suggest that a nuclear People's Republic threat to the Republic of China is likely, let alone imminent. However, the existence of Beijing's military capabilities, including nuclear ones, could deter Taipei from several kinds of otherwise feasible responses to low-level or even conventional military initiatives by the People's Republic.

12. For general assessments see, for example, Richard H. Yang, ed., *SCPS Yearbook on PLA Affairs 1987*, and *SCPS Yearbook on PLA Affairs 1988* (forthcoming), Sun Yat-sen Center for Policy Studies, National Sun Yat-sen University, Kaohsiung, Republic of China, 1988; Yu-ming Shaw, "A View from Taipei," *Foreign Affairs*, Summer 1985, especially p. 1056; Thomas W. Robinson, *Does Chinese Mainland Military Modernization Threaten Taiwan's Security?* (Washington, D.C.: American Enterprise Institute, no date, xerox). For naval plans, see also the *Washington Post*, 19 February 1989.

13. See *The Military Balance 1988–1989* (London: International Institute for Strategic Studies, 1988), p. 150.

14. See, for example, Martin L. Lasater, *Beijing's Blockade Threat to Taiwan* (Washington, D.C.: Heritage Foundation, 1986). See also June Teufel Dreyer, "Mainland China and Taiwan: Strategic Interaction," paper presented at a conference on Mainland China and Taiwan, George Washington University, 15–16 September 1989, xerox.

15. *The Military Balance 1988–1989*, p. 178.

16. In September 1989 Mikhail Gorbachev invited CCP boss Jiang Zemin to visit Moscow (*Australian*, 13 September 1989, p. 6). It is not surprising that the Taipei Ministry of Economic Affairs should have suggested to Republic of China businessmen that they should indirectly participate in Siberian development. *Free China Journal*, 25 May 1989, p. 8; also 18 May 1989, p. 1. That might give the Russians, too, a stake in Taiwan's welfare.

17. See some comments from Martin Lasater in the *Free China Journal,* 13 April 1989.

18. Cf. Shaw, "A View from Taipei," p. 1051.

19. Cf. D. V. Hickey, "US Economic, Political and Strategic Interests in Taiwan: The Ties that Bind," *Issues and Studies,* November 1986, pp. 64–66.

20. Thomas B. Gold, comments on "US-Taiwan Security Issues," unpublished paper presented to Washington Institute seminar, 26 November 1988, xerox.

21. Shaw, "A View from Taipei," p. 1051.

22. See, for example, *Free China Journal,* 10 August 1989, p. 1, and 14 August 1989, p. 5.

23. For example, the Republic of China's Mitac International Corporation has joined with General Electric of the United States to produce advanced electronics and information processing equipment. *Free China Journal,* 14 August 1989, p. 7.

24. For comments on the upgrading of weaponry and equipment as the Republic of China looks towards the 1990s, see the *Far Eastern Economic Review,* 30 July 1987, pp. 15–18, and Dreyer, "Mainland China and Taiwan."

25. The figures that follow are taken from the International Institute for Strategic Studies, *The Military Balance 1988–1989.*

26. One observer has suggested that the People's Republic would need some forty divisions for any successful invasion of Taiwan. Quoted in D. V. Hickey, "America's Two-Point Policy and the Future of Taiwan," *Asian Survey,* August 1988, p. 885. Their probable losses in any invasion of Taiwan would make the People's Republic vulnerable vis-à-vis third parties.

27. *The Military Balance 1988–1989,* p. 178; Dreyer, "Mainland China and Taiwan."

28. Foreign Broadcast Information Service, China (hereafter FBIS-CHI), 6 December 1988, p. 70; *Free China Journal,* 7 September 1989, p. 1.

29. FBIS-CHI 12 April 1989, p. 75; *Free China Journal,* 14 August 1989, p. 2.

30. Cf. Robinson, "Does Chinese Mainland Military Modernization Threaten Taiwan's Security?," pp. 2–3.

31. For a general review of the People's Republic–Republic of China relationship, see Lee Lai To, "The PRC and Taiwan—Moving towards a More Realistic Relationship," in R. A.

Scalapino, S. Sato, J. Wanandi, and S. Han, eds., *Asian Security Issues: Regional and Global,* Institute of East Asian Studies, Research Papers and Policy Studies no. 26 (Berkeley: University of California Press, 1988), pp. 165–95.

32. Cf. Frank Hsiao and Lawrence R. Sullivan, "The Politics of Reunification: Beijing's Initiative on Taiwan," *Asian Survey,* August 1980.

33. Cf. *People's Daily,* 1 January 1979.

34. *China Handbook 1984* (Hong Kong: Ta Kung Pao, 1984).

35. Deng Xiaoping, *Building Socialism with Special Chinese Characteristics* (Hong Kong: Joint Publishing Co., 1985), pp. 27–29.

36. For a longer-term and historical assessment of Beijing-Taipei relations, see Hungdah Chiu, ed., *China and the Question of Taiwan* (New York: Praeger, 1973). For a more recent view, see John Quansheng Zhao, "An Analysis of Unification, the PRC Perspective," *Asian Survey,* October 1983.

37. See, for example, *Free China Journal,* 7 September 1989, p. 2.

38. See, for instance, Thomas B. Gold, "The Status Quo Is Not Static," *Asian Survey,* March 1987, pp. 300–315. And Peter Kien-hong Yu, in "Relations between Peking, Washington and Taipei: Maintaining the Forbidden Triad or Building Fayol's Bridge," *Issues and Studies,* April 1987, pp. 86–110, argues, among other things, that the People's Republic is preparing to build bridges to the Republic of China by 1997 at the latest, with relations and communications through Hong Kong. Hong Kong would keep its trade agreements with Taiwan, and existing offices would remain open in both Hong Kong and Taipei, even after Hong Kong's return to mainland control.

39. See *Free China Journal,* 11 May 1989, p. 1, and 15 May 1989, pp. 1–2; also the *Australian,* 11 May 1989, p. 8.

40. Cf. *Free China Journal,* 18 May 1989, p. 1, and 28 August 1989, p. 1.

41. Cf. *Free China Journal,* 11 May 1989, p. 1, for some indications of Republic of China openings to the People's Republic, including family visits. Also the *Economist,* 1 July 1989, pp. 20–21.

42. It is sometimes said that this includes having learned, back home, the art and business of just how to bribe officials.

43. See Peter Kien-hong Yu, "The PLA Navy," in Yang, *SCPS Yearbook on PLA Affairs 1987.*

44. It is not entirely clear what these hints mean. It may mean less Soviet backing for Vietnam's interests in Kampuchea. But there have been no signs of any Soviet willingness to unilaterally abandon Cam Ranh Bay and the other major Soviet facilities. Whether the Soviets could announce plans to give up Cam Ranh Bay at some sensitive time in future—for instance at some critical stage of any future negotiations between the United States and the Philippines about Clark Field and Subic Bay Naval Base—remains to be seen.

45. If a carrier were built, it would be certain to be the first of at least two or three, given the problems of keeping carriers and their crews on station. Such policies would also necessarily imply the construction of carrier task forces, and such a development could significantly affect the balance of air power over the Taiwan Straits. However, any such PRC policy would be a very long-term one.

46. The excitement in Taipei, as elsewhere, was palpable. See *Free China Journal*, 25 May 1989, p. 1, for the text of the Government Information Office statement.

47. During the crisis, it seems that over a hundred Chinese military leaders signed a letter to the *People's Daily* that was never published. The letter argued against the use of force in Tiananmen Square and said, "The People's Liberation Army belongs to the people. It cannot confront the people, even more it cannot suppress the people and will never shoot the people." Quoted in Claire Hollingworth, "Yang's Political Commissars Back in Authority," *Pacific Defence Reporter*, July 1989, pp. 16–17.

48. *International Herald Tribune*, 4 September 1989, p. 5.

49. *Asian Wall Street Journal*, 11 July 1989, pp. 1, 20; *Economist*, 24 June 1989. However, Guangdong Province continues to have some two million people working for firms in Hong Kong, at about one-eighth of Hong Kong's wages, and in July one Guangdong town, Henggang, actually agreed to lease 60 million square feet of land for twenty-five years to an American company, with subletting permitted. *Economist*, 19 August 1989, p. 16.

50. In September, for instance, China borrowed $200 million abroad—but the interest rate was 0.75 percent higher than the

London Interbank-offered rate. *Australian,* 13 September 1989.

51. *Economist,* 1 July 1989, p. 19.
52. Among other things, the Beijing events have thrown into stark relief not just the depth of conflicts within the country, the PLA, and the government, but the incompetence of some ruling groups. That those who ordered and organized the repression in Beijing should have allowed the entire affair to be covered by the world's television cameras passes comprehension.
53. For a PRC view, see Chen Qimao, "The Taiwan Issue and Sino-US Relations," *Asian Survey,* November 1987, pp. 1161–75.
54. Liu Chih-kung, "Taipei-Washington-Peking Relations," *Issues and Studies,* July 1986, p. 17.
55. *Public Papers of the Presidents of the United States: Ronald Reagan,* 1983, vol. 1 (Washington, D.C.: Government Printing Office, 1984), p. 453. More generally, see, for instance, Hickey, "US Economic, Political and Strategic Interests in Taiwan: The Ties that Bind," pp. 64–66; Hickey, "America's Two-Point Policy and the Future of Taiwan," pp. 881–96; Taifa Yu, "The Reagan Administration's Management of the Taipei-Washington-Peking Triangular Relationship," *Issues and Studies,* November 1987, pp. 69–95.
56. Gold, comments on "US-Taiwan Security Issues."
57. Liu, "Taipei-Washington-Peking Relations," p. 19.
58. See, for example, C. L. Chou, "Dilemmas in China's Reunification Policy toward Taiwan," *Asian Survey,* April 1986, pp. 467–82, and Lucian W. Pye, "Taiwan's Development and Its Implications for Beijing and Washington," *Asian Survey,* June 1986, pp. 611–29.
59. Cf. Liu, "Taipei-Washington-Peking Relations," pp. 23–24.
60. See, for example, Roger W. Sullivan, "US Military Sales to China," *China Business Review,* March–April 1986, pp. 6–9; Denis V. Hickey, "US Arms Sales to Taiwan," *Asian Survey,* December 1986, pp. 1324–36.
61. Yu, "The Reagan Administration's Management of the Taipei-Washington-Peking Triangular Relationship," pp. 69–95.
62. Ibid., pp. 71–72.
63. Hickey, "US Arms Sales to Taiwan," p. 1324.

64. Ibid., pp. 1330–31.
65. Shaw, "A View from Taipei," p. 1051.
66. Yu, "The Reagan Administration's Management of the Taipei-Washington-Peking Triangular Relationship," p. 85.

SIX

United States–Vietnam Security Issues

Douglas Pike

A Strategic Overview

Neither Vietnam nor the Southeast Asia of which it is part commands the strategic importance in Washington of other geographic areas—Northeast Asia, for instance. The region has, in the last few years, taken on some new importance, chiefly for economic reasons: the United States trade balance has swung from Europe to Asia, there has been increased American investment locally, and there are new natural resource acquisition needs.[1]

The chief United States national interest that has flowed from such economic factors is in terms of access to the region and freedom to traverse it. Credible challenge by Vietnam to these interests is low. It is hard to envision a scenario in which Hanoi acting on its own, or even as a Soviet surrogate, would deliberately and effectively block American access or right of passage. Impaired access or uncertain passage due to instability, regional war, or generally chaotic local conditions implies a United States national interest of regional stability, or more precisely put, equilibrium. In the nineteenth century this condition was commonly termed "balance of power," in the twentieth century "ideological balance of power." Something of this original "balance" notion still obtains, but now it seems better described as a search for equilibrium.[2] It is, I would argue, in the interests of *all* nations of the Pacific Basin that *no* single nation dominate. Further, should any one nation threaten to dominate, it is in the interests of the rest to stand in opposition.[3]

American national interest with respect to Vietnam and in the broader reaches of Asia (as this is being written in October 1989) is undergoing thorough review in Washington as a result of recent emergent trends and new developments in a region that has always been highly dynamic. This lends a certain tentativeness to anticipating future American policies and actions with respect to Vietnam. However, there are transcendent American interests that do not change, even if new personnel appear on the scene, and it is to these that this paper is directed.

From the United States point of view there now exists a fairly stable equilibrium in Asia, which also applies to Vietnam, although admittedly relations there are minimal. I would argue that this stability is primarily a product of the *way* the Vietnam War ended. Put differently, if one surveys the Southeast Asian scene today in terms of United States national interests, one must conclude that generally speaking what is seen is satisfactory. This condition resulted from the outcome of the Vietnam War. Had the United States "won" the war, the scene in Southeast Asia today would be far less favorable to American interests. The same condition is applicable to Northeast Asia, namely, the *way* the Korean War ended resulted in generally satisfactory Northeast Asia regional conditions in terms of United States national interests, and had the United States "won" the Korean War the American geopolitical posture today would be far less satisfactory.[4]

Recent "negative" developments in Asia and particularly in Vietnam, including Soviet naval and aerial basing operations at Cam Ranh Bay and Da Nang, could affect this fundamental equilibrium, although Soviet activity to date appears more symbolic than strategic. In taking the measure here it is important to use an appropriate yardstick. It is misleading to evaluate the geopolitics of Southeast Asia in terms of traditional balance of forces, which is still necessary with NATO–Warsaw Pact considerations. Rather, the central fact of the United States–Soviet relationship in the area is regional competition.

One of the results of the Vietnam War was the emergence of a new triangular geopolitical relationship in Southeast Asia among the United States, the Soviet Union, and China and a certain level of struggle for regional influence and power. This is a normal condition, inevitable in any case, and not particular cause for alarm. As this triangular relation develops, the United States may have to be something of a guarantor of ASEAN interests at the superpower level; in effect, it must assure the ASEAN nations that Soviet naval presence is a normal, legitimate activity of a major power acting in a benign way (if in fact this is the case). Even now ASEAN is quietly pressuring Washington to "keep Moscow off our backs," as it has been phrased by one representative. At the same time, the United States may find it necessary to disabuse ASEAN leaders of their belief that China is their implacable enemy.

There is little to be said about national security issues—strictly or narrowly defined—standing between the United States and Vietnam. Neither poses much of a credible threat to the other, and such threat as might be perceived by either is indirect—that is, directed at their respective allies or associates. There are, it is true, areas of contention between the two. However, the traditional or orthodox definition of security has been vastly broadened in the past decade or so. After World War II American thinking about Vietnam turned almost entirely on national security considerations. Now that monopoly of military domination has been broken, semi-security issues have moved to the forefront. These include the traditional concerns for trade and aid and newer considerations such as technology transfer, multinational corporation interdependence, nuclear proliferation, and arms transfers. Increasingly, it seems these quasi-security factors will influence United States foreign policymaking with respect to Asia as well as to Vietnam.

It also seems probable that as we move toward the next century American policymakers, partly by design and partly in reaction to unfolding events, will increasingly be guided by the "equilibrium thesis" as the overarching principle of United States foreign policy with respect to Vietnam, Southeast Asia, and Asia as a whole. They will work for a condition of sociopolitical, economic, and strategic equilibrium within the framework of competing regional institutions. Probably this will consist institutionally of a matrix of organizations, some large and some small, some with broad general purpose and some with narrow specific objectives, some governmental, some private, some multinational. It will be a vast, organized arena in which a contained struggle for power will be conducted. Within this arena will be contained the United States–Vietnamese relationship.

Early Relations

American national interest in Vietnam and Asia began early in the nineteenth century and became serious about the turn of the century with the American thrust into the Pacific in the name of presence and mission. From the start, the American interest was security-oriented, whether a search for coaling stations or the march of imperialism into the Philippines. Following World War I the policy was to fix and maintain a Pacific balance of power, as with the 1920s conference to limit naval construction. Out of World War II came the "Munich syndrome," the notion that isolationism had triggered the war and that the lesson to be learned was that aggression not stopped early had to be stopped later at higher cost. This became an influential concept among policy makers and was largely the reason for United States entry into both the Korean and Vietnamese Wars.

United States geopolitical thinking about Indochina/Vietnam thus inherited elements of the "balance" notion and the "Munich syndrome," to which was added the ideological challenge of Marxism-Leninism. The Soviet Union (and China after 1949) were seen as potential aggressors. To the extent that there was any thinking about Indochina/Vietnam in Washington it was in terms of the threatening "red tide" represented by the Soviet-Chinese alliance. In Asia this meant that both geopolitical and ideological imbalances had to be redressed. French efforts to preserve their Indochina empire were seen as a test of this policy. It was widely believed—and not only by the simple-minded embracing a mechanistic "domino theory"—that if Indochina fell, soon would follow most if not all of Southeast Asia. Hence, even against the better judgment of American leaders, France was supported in a cause that few thought it could win, but which required a United States effort in order to help restore the balance.[5] A two-Vietnam arrangement was salvaged from the debacle, a measure that the United States was then obliged to perpetuate. By 1965 the United States had been drawn into full intervention in South Vietnam, defining victory at the time as keeping South Vietnam out of the hands of international communism. When the Sino-Soviet alliance dissolved (or was revealed never actually to have existed) the major rationale for American involvement in Vietnam evaporated. The

American purpose in Vietnam then became disengagement under the best possible conditions (the sardonic comment at the time had it, America is fighting in Vietnam to get out). Where once it had been imperative that the communist advance be stopped, now this idea no longer seemed very important.

The "relationship" of the Democratic Republic of Vietnam (DRV) or North Vietnam to the United States during the war years, in Hanoi Politburo terms, was a by-product of its single, great golden cause: unification of North and South Vietnam. The United States after 1965 was regarded as the chief obstacle to achieving this goal. That fact conditioned the leadership's thinking about Washington then and now.

The leaders in Hanoi devised a strategy against the Americans to achieve unification. In a sense, they continue to use it today in pursuit of all current national interests. The wartime strategy was called *dau tranh* (struggle). It was a mix of armed struggle and political struggle in a communicational context.[6] Political *dau tranh* helped nullify American military strength (inhibiting the bombing campaign, for instance) and discouraged United States determination (through prolongation of the war). It was an effort to co-opt the Americans—that is, to force and maneuver them into furthering the North Vietnamese cause of unification. Armed *dau tranh* tried to drive them out militarily but was less successful. Eventually the combination delivered, however, victory and unification achieved.

Thus ended a Hanoi–United States "relationship" that had begun in 1945 and had continued unchanged and hostile for thirty years. It had been a long, deadly embrace, but it suddenly was replaced by nothingness. Where there had been preoccupation, each with the other, now there was only a void.

Postwar Relations

The history of the United States–North Vietnam association since 1975 is one of studied indifference on both sides. For the Americans it has been a dramatic transformation, from being a nation mesmerized by Vietnam to becoming one that seemed hardly aware Vietnam existed. It was not simply indifference, however. Rather, on both sides attitudes developed carrying strong psychological overtones, nourished by the dark, untapped parts of both American and Vietnamese psyches. For the Americans there has been the bitter aftertaste of a long and unsuccessful war. For the victorious Vietnamese there has been a strange descent into a kind of disorientation.

In terms of theoretical policy constructs, the United States facing Vietnam in 1975 had three policy options. It could mount an effort to roll back communism in Indochina by funding resistance movements or possibly again intervening (a choice made politically impossible by American public opinion); it could actively seek to contain Vietnamese expansionism (a choice rendered politically impossible for the same reason, but one that, except for Cambodia, became unnecessary); or it could establish a minimal formal diplomatic relationship with the tacit understanding that it would gradually develop over the years (a choice the United States was on the verge of making before backing away). The United States chose none of these. Instead, it opted for what might be called a holding pattern in lieu of policy. There would be a "dust settling" interim, during which United States policy would be to follow the lead of ASEAN/China, in effect allowing American associates in Asia to fix the parameters of the Washington-Hanoi relationship. Thus, it became a kind of non-relationship. There was little pressure on Washington to move out of such a holding operation, and Vietnam's 1979 invasion of Cambodia provided an irrefutable argument for continuation of the non-relationship (at least until the invasion was ended and Vietnamese troops withdrawn).

For their part, the Hanoi leaders in the first year after the war ended set about devising a diplomatic strategy rooted in the *dau tranh* concept, to force from Washington part or all of the $3.25 billion they claimed the United States owed them.[7] Their calculation, and this appears still to be their thinking, was that by various "struggle" tactics—doling out POW remains, vilifying the United States in international forums, appealing to American war-guilt feelings—eventually they would succeed. Should anyone suggest that theirs was a hopeless pursuit, these leaders would reply that the same thing had been said to them when they began war with the United States. Hanoi's subsequent dealings with the Americans have been consistently characterized by this diplomatic "protracted conflict" approach. The leadership clearly is playing for the long haul and is uninterested in any diplomatic "quick fixes."

In the first months after it took office, the Carter administration made a serious effort to persuade the Vietnamese leaders to establish normal relations, defined as an exchange of embassies, without conditions. It dispatched to Hanoi the Woodcock Mission (named after its chairman, Leonard Woodcock) to explore Politburo thinking. The Hanoi leaders took a "struggle line" approach. They set conditions; they spoke of American economic "obligations"; they mentioned the $3.25 billion figure. The Hanoi press made reference to "war reparations." The Americans explained that the United States foreign aid process required congressional authorization and involved the domestic politics part of the democratic process. They suggested that embassies be exchanged and that the newly arrived Vietnamese ambassador to Washington solicit economic aid through representations at the Department of State and by lobbying on Capitol Hill, since that is the way it is done. The Politburo, however, stood by its "precondition" position: aid before recognition. The Americans demurred, and the mission ended inconclusively.

There the matter stood for the next year or so—marked by occasional UN and deputy-assistant level talks at the UN and in Paris, during which the United States acquiesced in UN membership for Vietnam and pledged to end trade restrictions once diplomatic relations were established. This was a dynamic period, however. Vietnam-China relations were deteriorating and finally reached the point where Hanoi officials became sufficiently fearful of China to drop their precondition on United States relations. But at the same time United States–Chinese relations were warming— it was the time of "opening to China." Increasingly, the Carter administration found the matter came down to choosing between Vietnam and China—no hard choice to make. Washington took no action on the signals from Hanoi in the fall of 1978. When, on Christmas Day 1978, the Vietnamese invaded Kampuchea, recognition was rendered impossible. Thus in December of 1978 Vietnam was denied what almost certainly it could have had in January—and would have had but for poor Politburo judgment.

As it turned out, the United States fortuitously "dodged the bullet" since it would have been embarrassing, to say the least, for an American ambassador to arrive in Hanoi at about the time

Vietnam was signing a treaty of friendship with the Soviet Union, joining CMEA, and invading Cambodia.

The Reagan administration fleshed out the policy of benign neglect toward Vietnam that had marked the last of the Carter years and developed the derivative policy of following the ASEAN/China lead with respect to Vietnam. There was much to recommend this approach. Since the Nixon Doctrine years the United States had been trying to get ASEAN to assume more responsibilities in regional security. However, this was not a policy so much as a holding operation that by its nature automatically meant that the United States abrogated a leadership role in Southeast Asian affairs. It was not a feasible long-run policy.

Within the Reagan administration there were varying opinions on the wisdom of the nonrecognition policy. The most supportive line was taken by the State Department, principally because recognition was seen as disruptive for United States–ASEAN and United States–Chinese relations. The most critical position was found on Capitol Hill, where a number of Republican congressmen forthrightly advocated American recognition of Vietnam. The congressional scene, however, was complicated by cross-purposes, chiefly domestic political factors impinging on foreign relations considerations. The Pentagon position, perhaps somewhat unexpectedly, fell between that of the State Department and that of the Hill. The rationale employed by those in the Pentagon favoring recognition was that it could ameliorate Soviet presence in Indochina. The White House tended to treat the issue strictly in terms of domestic politics, relating it to resolution-of-casualties issues and émigré Indochinese politics.

Public opinion in the United States tended to mirror this spread of perception in Washington. There were spirited differences of viewpoint, especially in American academia. Single-interest groups formed to advocate or oppose United States diplomatic recognition of Vietnam and became part of the general political process. The division here was along traditional liberal-conservative lines, although there were numerous crossovers—that is, conservatives advocating recognition as a means of inducing Hanoi to account for American MIAs and liberals who wanted to make Hanoi pay for its postwar aggression. In the late 1970s, elements of the American business community—spearheaded by the American Chamber of Commerce in Hong Kong—pressed for United States recognition of Hanoi. This pressure evaporated after the breach of Sino-Vietnamese relations, when Beijing told the businessmen to choose between China and Vietnam, and sensibly most of them chose China. The antiwar activists split after the war over the Vietnam human rights issue (boat people, re-education camps, new economic zones) and over causes of the holocaust in Kampuchea. Finally, there was the rise of a new United States pressure group, the émigré Vietnamese. These now number about 900,000, and while most of them remain apolitical, they are becoming increasingly organized and generally oppose recognition. American public opinion remains divided, with only a minority favoring United States relations with Vietnam, the remainder indifferent or opposed. A decision to recognize Hanoi therefore would probably draw no particularly strong or sustained reaction from the country, except from the Vietnamese émigré community.

Outstanding Issues

National interest issues stand between Vietnam and the United States and will continue to do so regardless of whether or not formal diplomatic relations are established. Generally speaking, it is probably in the interests of both that these problems be resolved. The basic purpose of diplomacy is to resolve such outstanding issues or, if this is not possible, to ensure that the other side understands the reasoning involved. Obviously, maintenance of formal diplomatic relations facilitates this process, although of course problems can be addressed without such a mechanism.

In any event several questions will continue to be central to United States national interest with respect to Vietnam.

The first issue is the peace process in Cambodia. This sad, bloodied little country represents the major contention between the United States and Vietnam. It is also the touchstone of ASEAN and Chinese policy with respect to Vietnam. The Reagan administration's position was that there could be no formal relationship with Vietnam until its troops were gone from Cambodia and a comprehensive peace settlement was arranged. Vietnam has asserted that as of the end of September 1989, all of its military forces have left Cambodia, although this has been challenged by the Chinese and the Coalition Government of Democratic Kampuchea (CGDK) on the grounds that several hundred thousand Vietnamese "settlers" remain and are controlling the government in Phnom Penh.

Despite the current "peace process," which failed to progress at the Paris meeting in July–August 1989, the most likely prospect for Cambodia in the foreseeable future—that is, the next year or so—is simply more of the same. The struggle will go on without resolution or decisive development.

Many observers predict a "fading war" scenario, in which the Phnom Penh government gradually increases its viability and legitimacy at the expense of the resistance forces so that eventually opposition is reduced to troublesome but not strategically significant backcountry guerrilla activity, as exists in northern Malaysia and northern Burma.

A less likely prospect, but still a possibility, is the establishment of a new governing structure in Phnom Penh providing acceptable representation to the major contending parties, namely the CGDK (composed of the Khmer Rouge, the Sihanoukists and the Son Sann forces) and the Hanoi-backed People's Republic of Kampuchea (PRK). This is the "all-Khmer" solution (or "Khmer only," implying exclusion of Vietnam). Few outsiders realize what a monumental task is involved in putting together a Cambodian coalition government. The People's Army of Vietnam (PAVN) is not in the country for altruistic reasons, but its military government represents the only government below the provincial level. Precipitous withdrawal of PAVN without a new governing system simultaneously moving into place would throw the country into anarchy and a power struggle that would soon devolve into thirteenth-century warlordism, in which the suffering of the Khmer people would be even worse than at present.

Possibly the Phnom Penh government will be able to destroy the resistance and more or less "pacify" the country, or at least confine armed resistance to the more remote parts of the Cardamom mountains.

Efforts initiated at the Jakarta Informal Meeting (JIM) at Bogor, Indonesia, July 24–28, 1988, represent a serious effort by ASEAN to get the Cambodian problem off dead center of the agenda. There was motion, perhaps movement in this direction during 1988 and into 1989. At this writing, however, there has been little progress in settling the basic issue, the unresolved question, what the future governance of Cambodia is to be and how it is to be achieved. The issues are not PAVN pullout or preventing the return of Pol Pot per se (would that it were that simple) but how to install a new governing structure in Phnom Penh that is (a) acceptable to all of the major contending Khmer factions; (b) able to meet Hanoi's legitimate national security interest that a hostile regime not come to power in Phnom Penh; (c) capable of allowing the Chinese the presence, influence, and status that they feel is their due; (d) acceptable to ASEAN (particularly Thailand, in terms of its security needs); and (e) a solution that major outsiders, the Soviets, the United States, and Japan, can live with (a lesser problem). Within these general requirements remain the specific issues: a PAVN withdrawal timetable; cease-fire or end to resistance activity; recruitment (and funding) of outside peacekeepers; devising of the mechanism for provisional rule (presumably a coalition government); subsequent establishment of a new government and constitution based on free, supervised elections; disarming or sequestering of all military forces and creation of a new, single Khmer armed force; and effecting of a mechanism for enforcement or guarantee of the arrangement by outside governments.

The second major issue from the standpoint of the United States national interest is the presence in Indochina of the Soviet Union, which has established a series of air and naval facilities as part of a broader effort to augment its military prowess in the Pacific.[8] This represents a certain threat to the United States and its allies and associates in the region. The exact nature and degree of that threat is variously defined by analysts and observers. A common view influential in the United States and Europe is that Moscow's moves into Vietnam and the Pacific stem from a natural growing concern for a region that increasingly affects Soviet security and economic interests. Therefore Soviet presence in Vietnam is normal and essentially benign (or at least not deliberately aggressive). Standing opposite this rather sanguine view is the contention of some that Soviet military facilities in Vietnam constitute a limited strategic threat. Most analysts believe that Moscow's military planners concluded early that their Vietnam bases were highly vulnerable and would be unavailable in a war with the United States. Therefore they have not incorporated them into their war scenarios. Short of total war, however, the bases do have strategic utility.

The Soviet Union and the Socialist Republic of Vietnam (SRV) have a military alliance in all but name. They conduct combined defense planning, run joint training exercises, and are presumably prepared for combined military operations. Soviet ships and military planes make full use of Vietnamese facilities and appear to be granted any support they require. This Soviet-Vietnamese defense arrangement does constitute a strategic threat, but one that is essentially psychological and operant only in circumstances short of total war. In a condition of cold war (or limited war) the bases have great utility. They help encircle China and would be useful in any limited war involving the Soviets. The bases would be essential for Soviet Afghan-style intervention in the region. Their existence intimidates Asia, particularly Japan, with its vulnerable sea lanes.

Soviet naval and air facilities at Cam Ranh Bay and Da Nang represent the most overt evidence of the Soviet-Vietnamese alliance. From time to time ASEAN officials (encouraged by Moscow's spokesmen) suggest that the USSR "trade" its facilities in Indochina for the American military facilities at Clark Field and Subic Bay in the Philippines. It is hard to believe that in this Moscow is doing more than making mischief; in any event the idea has been denounced by Hanoi officials, who caustically declared that Cam Ranh Bay is not Moscow's to "trade."

I do not believe that the Soviet-Vietnamese association is as close or as durable as most observers contend. It is based on Soviet opportunism and Vietnamese dependency (for food and weapons) and will last only as long as the USSR considers it useful and, for Hanoi's part, only as long as Vietnam is unable to feed itself and until the China threat subsides. In any event, I do not believe that a nominal change of United States–Vietnam relations, as in the establishment of diplomatic relations, would have any important effect at all on the Soviet-Vietnamese alliance.

The third issue for the United States is regionalism. Probably this is the most important question between the United States and Vietnam, since it has to do with the future political configuration of Southeast Asia in general and Indochina in particular.[9] As an organizing principle in international relations, regionalism clearly is a growing force throughout the world. The regional organization is seen as the best available mechanism to deal with multilateral international problems. In part, this is a reaction to the failure of the more ambitious international systems such as the

United Nations, and in part it is a matter of sheer necessity. In any event it seems clear that history is pushing the world in the direction of regional groupings.

It is my belief that regionalism is the coming major geopolitical force in Southeast Asia, including Indochina. There is of course regionalism and then there is regionalism—our regions and their regions: ASEAN and the Federation of Indochina. ASEAN as a system has come far—and has far to go. The regionalism of Indochina is moving, albeit very slowly, in the direction of being dominated by Hanoi. It seems probable that by early in the next century a Federation of Indochina dominated by Hanoi will have come into formal existence. There is nothing deterministic about this: it is simply that one cannot see on the horizon sufficient counterforce to prevent it. The influences of regionalism already have given impetus to solidifying and formalizing relations among Vietnam, Laos, and Cambodia, moving them from their long-established ambiguous "special relationship" to a much clearer "alliance" that Hanoi says now characterizes the association. It is the road to federation.

The United States as well as ASEAN and China eventually must come to grips with the question of the acceptable limits of Indochinese integration. It is a very difficult problem, which is why it has been dodged by all for so long and also why Cambodia is so important. Is a federated arrangement, or even a confederated arrangement, acceptable? And if not, what exactly are outsiders prepared to do to prevent it?

Another issue is the exodus of Indochinese, chiefly Vietnamese, from their homelands, the so-called "boat people" phenomenon. To a large extent, this defies solution, even given agreement by all parties involved as to what should be done, since the people involved are fleeing their homelands because they find conditions there intolerable, both politically and economically. The only sure way of halting the flow is to eliminate the cause. Hanoi officials have been trying to do this (at least with respect to the economic malaise) for nearly a decade now, without marked success.

A number of other bilateral questions of lesser magnitude than those discussed above stand between the United States and Vietnam. There are various economic matters—Vietnamese assets frozen in the United States, nationalized United States property, demands for indemnification by both sides. There are humanitarian problems involving divided families, the Orderly Departure Program, Amerasians and their families, and the re-education camp inmates seeking emigration.

Finally, there is the knotty, most difficult resolution-of-casualties issue—the need for an accounting by Hanoi, to the extent it can, for the fate of some 2,000 American military men of the Vietnam War era who are still listed as missing in action or as fate unknown. The MIA/POW problem cuts to the political bone in America because of its deep psychological meaning. It challenges the fundamental sense of responsibility of our highest officials. It influences Americans who fought the war and those who opposed it.

The problem is also singular, not one that usually appears on embassy agendas. Normally, and logically, nations treat the assuaging of bereavement as a humanitarian matter, not as something for their diplomats to negotiate. Further, while most issues are negotiable, this one is not. Since 1975 the basic American approach has been to try to convince Hanoi to put the resolution of casualties into a separate compartment and deal with it independently of all other foreign policy questions. For more than a decade the United States had no success in changing what amounted to a myopic, anachronistic mind-set on the part of the Vietnamese. In 1988, however, we saw a glimmer of a change in Hanoi's policy. The Vietnamese accepted the principle of selecting out certain "humanitarian issues" (including the resolution of casualties, Hanoi's need for specialized medical treatment for war casualties, wartime environmental damage to Vietnam, Amerasian children) and agreed to pursue these questions regardless of the ups and downs that might occur elsewhere in the bilateral relationship. To date Hanoi's behavior has adhered only inconsistently to this principle.

Vietnam Threat Potential

The chief Vietnamese threat to United States interests, to the extent that a true threat exists, derives from Hanoi's association with the Soviet Union. Vietnam does not represent a credible military threat to any American associate or friend in Southeast Asia except Thailand, since it does not have the air and naval power to project force over long distances.[10] PAVN could invade and occupy Thailand in a matter of days, although there are many compelling reasons for it not to do so, not the least of which is that it probably would find that its impasse in Cambodia had been extended to the entire Indochina peninsula. In a limited war with China, PAVN could probably hold its own for a lengthy period, but could not win ultimately.[11]

In addition to being a limited orthodox military threat to American allies in the region, Vietnam can also offer the indirect threat of funding and supporting insurgencies in the ASEAN countries. While these insurrections might not prove successful, with Vietnamese guidance and aid they would be troublesome and costly to suppress. A primary candidate here is the Philippines. There were persistent rumors in 1988 that Vietnamese weapons and military supplies were reaching the New People's Army in Luzon and possibly the Moro rebels in Mindanao. The reports were never confirmed.

Vietnamese ideologues writing theoretically about the future of the region say they expect that the ASEAN countries will eventually shift politically to the left, changing institutionally until they all become "people's republics." Official doctrine in Hanoi holds that the governments and societies of noncommunist Southeast Asia are illegitimate and transitory, soon to be swept "into the dust bin of history." They argue that the doctrinal problem for Vietnam is only tactical—how to push history along. The question for them is simply whether to organize and fund insurgencies and other left-wing causes or just to let history take its course.

For the present it appears Hanoi has ruled out aiding and abetting insurgencies. In the mid-1970s, PAVN generals systematically examined the region's guerrillas, concentrating on the three groups in Thailand, and concluded that the rebels did not have the qualities that promised success. This policy of ignoring insurgent appeals for assistance may change, of course, but clearly Hanoi must be convinced that a rebel force has real prospects before it will back it.

The third threat that Hanoi could offer American interests in the region can be labeled politico-institutional, and falls in that gray area between insurgent war and politics. The concept surfaced in late 1975 when Hanoi, flushed with victory, began looking for new worlds to conquer. Party theoreticians mapped out a national security strategy for use in Southeast Asia. Vietnam would define regional peace as an end to economic cooperation with Western and Japanese capitalism. It would offer a supporting doctrinal rationalization, which would be a synthesis of nationalism and collectivism. Its slogan would be Hanoi's Vietnam War slogan: Forward under the banner of national independence and socialism. The doctrine would say to the region that a nation cannot be truly independent and still have ties with external capitalism. Nor can it be neutral in the great struggle between capitalism and socialism. However, choosing socialism as an economic system does not mean embracing world socialism; nor does it require siding with either of the two great socialist superpowers. It requires only that a nation be "non-aligned" economically, the central enemy being not capitalism but economic interdependence.

In Hanoi's calculation, the strategic benefit to be gained would be to force the United States to cope with a Vietnam-led collaboration of economic saboteurs determined to frustrate American economic ventures and, if possible, shut the United States out of the region entirely.

It was a vastly ambitious scheme, first floated at the 1976 Non-Aligned Nations Conference in Colombo. After Vietnam's time of troubles began in 1979, little more was heard of the idea in communist theoretical journals; but it is still alive in Hanoi, and we may not have heard the last of it.

In evaluating Vietnam in national security terms, it is important neither to understate nor to exaggerate its threat potential. Focus should always remain on Vietnam as an associate of the Soviet Union. Moscow should be continually reminded, by the United States as well as by the ASEAN countries, that it is held accountable for the use to which Vietnam puts the weaponry the Soviets supply, and that in inhibiting military adventurism by Hanoi the Soviet Union serves its own national security interests.

Hanoi Looks at Washington

If asked to characterize present relations with the United States, the leaders in Hanoi would probably describe them as singular (unlike those with any other country), minimal, highly charged psychologically, and involving unique contentious issues. Most

Americans would regard that description as a mirror of their view of Vietnamese relations. Just as the "Vietnam experience" is deeply embedded in the American consciousness, manifesting itself in a hundred rational and sometimes irrational ways, so the "American experience" is deeply embedded in the Vietnamese consciousness.

That consciousness can be evaluated at three separate levels: historical or philosophical; as influence on Vietnamese domestic politics and on economic, diplomatic, and national security interests; and strategic/tactical.

The official party line in Hanoi with respect to the Americans and the Vietnam War is that Hanoi wants to forgive and forget, a contention that should be evaluated in light of Vietnamese cultural values, which are that the Vietnamese never forget and seldom forgive. However, exactly what the Vietnamese do think about Americans today cannot be fully established since we do not have the access needed for academic research. Based on anecdotal data it seems clear that something of a mirror image to American attitudes does exist.

Some Vietnamese, particularly those directly and adversely affected by the war, bitterly hate the Americans and probably always will. A larger number, found mostly in the South, remember the Americans with a certain fondness, some even entertaining hopes that somehow eventually the Americans will put things right. Many, probably the majority (included here is that half of the population under the age of eighteen), hold views that are neither salient nor particularly unfriendly, despite the agit-prop campaigns directed against them by the party since childhood. A similar division of attitudes, with some slight adjustment, perhaps, would fit American opinion of Vietnam.

While the historical and philosophical attitude toward the United States may prove to be most important in the long run, and while the third level (strategy/tactics in the pursuit of national interest) logically should be Hanoi's imperative consideration, it is the second dimension (internal political infighting) that dominates relations at the moment. The scene in the Politburo in Hanoi as this is being written is one of intense political struggle with no indication of an immediate resolution. The generational leadership succession process has long been under way and continues, but at glacial speed. The leadership is preoccupied by domestic problems, chiefly economic. Secondary policy questions, such as future relations with the United States, are shunted aside, or more precisely, put on hold. This is not a permanent condition, of course, but one that probably will continue for the next year or so.

Policymaking at the Hanoi Politburo level is essentially Leninist-style collectivism. All decisions must be minimally acceptable to all thirteen Politburo members (not out of considerations of ego but because of the constituencies that members represent and are beholden to). Day-to-day political activity at this level is Chinese-style politics of factionalism. The two major factions at present are labeled by outsiders (chiefly for ease of reference) as reformers and conservatives (or neo-conservatives). Political infighting in Leninist systems with a heritage of Confucianism is not nakedly personalized—that is, the politics of entourage—but cloaked in doctrine. The battle is between factions over doctrinal issues, one of which is how best to deal with the United States. There is agreement between the reformers and the conservatives that what Vietnam wants and needs from the United States is service to Vietnam's national security and economic development assistance. Differences of opinion are not over ends but means. The conservatives' view is that the best way to deal with the Americans is through *dau tranh,* sustained, prolonged application of pressure—to "stonewall" them—on the grounds that everything Hanoi has gotten from the United States has come through such protracted "conflict." The countervailing view of the reformers is that, while this once was true, it is no longer. The reformers hold that now a softer, more forthcoming approach is appropriate. The present division on this issue at the Politburo appears to be eight to three against, with the three military members having no particular position (possibly the count is eight to six).

There is no intent here to convey the idea that the political scene in Hanoi is self-destructive. On the question of United States relations, as on all other policy issues, so far as we can determine, the leadership remains united and in basic agreement. Whatever the "vote" might be in the Politburo on this or any other matter, what is of overriding importance to them is that the Leninist-style collective government operational code of Politburo decisionmaking continues.

Until the factional political infighting settles down in Hanoi and until the generational transfer of political power is completed (and of course the two are intricately interrelated) we must expect the

Politburo to temporize and procrastinate in all policymaking. Such has been the pattern since the death of Le Duan (1986). During 1988 the world witnessed an almost farcical performance by SRV Foreign Minister Nguyen Co Thach and other Hanoi representatives in dealing with the United States—backing and filling on the Vessey mission and on-again, off-again policy positions on MIA searches, re-education camp inmate emigration, the Orderly Departure Program, and the Amerasian children issue. It was what might be called a condition of self-imposed indecisiveness, mandated by the Politburo leadership. It has required them to deal with foreign representatives in a puzzling manner, marked by a great deal of vagueness and no small amount of dissembling. In a typical example, a Vietnamese official is approached by a foreign representative who seeks to sound him out on some subjects such as peace in Cambodia, recognition of the United States, rapprochement with China, economic intercourse, distancing Vietnam from the Soviet Union, ending Vietnam's diplomatic isolation. The Hanoi official may privately support the policy suggestion his visitor proffers. Probably he is a reformer (the conservatives have far less to do with foreign visitors) and may even believe that eventually Vietnam will go in the direction suggested by the foreigner. However, he knows that any change of policy at the moment is politically impossible. Hence his task is to turn the suggestion in a way that does not alienate his visitor, keeps his hopes alive, and leaves him convinced that eventually Vietnam will in fact accept his suggestion. This treatment characterizes the question of relations with the United States. When questioned point-blank by visiting journalists, Vietnamese leaders are obliged to sound forthcoming, as required by Confucian courtesy if nothing else.

The third level of Vietnamese perception of the United States, what might be called orthodox or standard, is the Politburo's view in terms of its two major national interests—Vietnam's security and Vietnamese economic development. These have been Hanoi's two major national goals since 1975 and, although they have been ineffectively pursued by the leadership, they remain the central intent. Future leaders can be expected to evaluate relations with Washington by asking: will the association enhance Vietnamese security (or at least not endanger it) and will it contribute to economic nation building?

It would seem that the answers to both questions are self-evident, that both interests would be served, even if only modestly, by a closer relationship. However, as viewed from Hanoi, particularly by the conservatives, the matter of United States relations is a complex one, difficult to evaluate and fraught with potential risk. The conservatives, perhaps reasonably, can question whether closer United States relations would (or could) enhance Vietnamese national security. Vietnam's chief need is for war material, virtually all of which must be imported since there are no arms factories in Vietnam. The United States could hardly be expected to ever be in a position to supply Vietnam with weapons for pursuing its war in Cambodia or for use along the China border. The best that might be hoped for from a relationship would be that it could deter the United States from adding to Vietnam's national security problems.

With respect to future United States–Vietnamese economic relations, Hanoi's interest obviously is in American economic assistance. The question the Politburo asks is not about need but about availability. Vietnam today is beset by massive economic problems resulting from years of neglect and more years of failed remedial measures. The economy is stagnant and anachronistic, with a narrow infrastructure. Vietnam is one of the poorest nations on earth. It is primarily agricultural, and what industry does exist (in the North) is antiquated and unproductive. It has one of the world's worst trade deficits. Hence by all logic the leadership should solicit American economic assistance virtually at all costs. However, the same assessment the Politburo makes as to the probable willingness of the United States to serve its national security needs, also applies to potential American economic aid. Vietnam would like massive assistance in the form of congressionally appropriated economic aid, something that they judge unlikely, and that probably is. Some private American economic assistance, chiefly remittances from émigrés, does reach Vietnam, despite legal prohibitions by United States law on such transfers. This has been estimated to amount to at least $100 million a year.[12]

Trade with the United States is of course possible, but prospects are unpromising for the near future. The fact is, Vietnam has little to sell and little hard currency with which to buy. This is not likely to change until its economic infrastructure can be developed. Even then, trade will largely consist of commodities such as coal, lumber, seafood, rubber, and spices.[13] What Vietnam could use, if the United States were willing to supply it, is technology transfer—that is, assistance in minerals exploitation, rationalizing the agricultural sector, and supplying the special tools and managerial technology that enable modern administrators to function efficiently. Technical assistance—data, advisers, systems—certainly will be a major factor in any future American negotiations with Vietnam.

In sum, it appears to be the judgment of the Hanoi Politburo that, although some economic benefit might accrue from closer relations with the United States, it would at best be marginal in the general scheme of things. Hence, economics does not act as much of a goad or a catalyst in Hanoi. If the United States were willing to supply Vietnam with significant economic aid, and if the Vietnamese leadership became convinced of this, probably this would bring a marked change in behavior. In the past, during the war and afterwards, economic considerations were never much of a factor in the Vietnamese leadership's thinking in either Hanoi or Saigon. In the future this may prove to be less true.

In any event, in Hanoi at the moment, the question is more doctrinal than economic. Internal logic would dictate that the Politburo should subordinate ideology, domestic political infighting, even perhaps national security to some extent, in order to meet the imperative demands of the economy. Given the present makeup of the Politburo a rational approach is highly unlikely. The conservatives remain in control. Earlier these conservatives (whose faction was then labelled the ideologues, vis-à-vis the pragmatist faction) opposed allowing economic factors to become influential, using the rationale of maintaining ideological purity. The argument is no longer made. Now the case is made—as part of the broader doctrinal dispute over change and continuity—that introducing change in the economic sector, while perhaps in itself admirable and even necessary, risks inadvertent changes that would prove deleterious. The same argument is made across the board against all proposed policy changes. It would be well to withdraw PAVN from Cambodia, but will a hostile regime then come into power in Phnom Penh? It would be better if Vietnam could distance itself from the Soviet Union, but would this not risk destroying the relationship? Opening up the country to foreign investment would bring in foreign money, but also unmanageable foreign influence. The conservatives, in making these various cases, do not argue against change per se, only against reformist policy changes that do not fully take into consideration the ramifications that might be produced. Rather than opposing change, they caution against rapid change. They believe they can in this way ensure slow, safe change. The result, however, has been change at such glacial speed that it amounts to no change at all. For example, the conservative influence in the 1988 appointment of Do Muoi as prime minister (a man who can be described as the Andrei Gromyko of Vietnam) produced a case not of going slow but of moving backwards.

The Future

In its first year the Bush administration's basic policy attitude with respect to Hanoi appeared to be "normalizing" relations.[14] This goal is now to be held in abeyance, for a temporary but indeterminate period. This was not regarded as either a putative or an ideological position, simply a pragmatic one. Establishment of diplomatic relations is to be treated as a question of timing, to be pursued at whatever level seems appropriate when the proper conditions obtain. One necessary condition is PAVN withdrawal from Cambodia as Hanoi's contribution to the peace process. By mutual agreement, the matter of the resolution of casualties has been transferred to the "humanitarian" category, and is no longer officially classified as an "issue." Something of a semantic dodge is involved here, but the change is also something the United States has been trying for years to get Hanoi to accept.

It seems likely that, sometime during the Bush administration's tenure, formal American diplomatic relations with Hanoi will be established at the lowly interest-section level (though possibly at the full embassy level), beginning minimal diplomatic intercourse of the kind common among nations of the world. This would not mean that a new ambience had developed between us, or that either side had changed its opinion of the other, or even necessarily that United States economic aid would begin flowing to Vietnam or that Hanoi would open its POW files to us. It is important to understand that an exchange of representatives would not automatically benefit the United States. Advocates in

Washington over the years have tended to see diplomatic recognition as a cause-effect process. They would list American national interests—a diminution of Soviet influence in Indochina, reduction of Vietnamese intrusion into Cambodia and Laos, economic investment opportunities for American businesses—and then imply that these would follow more or less automatically once an American ambassador had arrived in Hanoi. Diplomatic recognition is no panacea for the problems between the two countries. This is not necessarily an argument against recognition, only counsel that representation is one thing, problem resolution another.

Conclusion

From this brief survey of United States–Vietnamese relations we can draw certain tentative conclusions:

- The United States "non-policy" with respect to Vietnam, though reasonable for a time while the dust settled in Southeast Asia, appears now to have run its course. The time has come for a new determination of relations, both by Washington and by Hanoi.
- A still unresolved leadership struggle in Hanoi over reforms and involving a generational transfer of political power, combined with serious internal economic difficulties, appears to stymie for the moment any change of Vietnamese relations with the United States. This is a transitory condition, one not expected to last for more than a year or so.
- The future American relationship with Vietnam, as with United States relations elsewhere in Southeast Asia, will not be as controlled by national security considerations as it was in the past. Rather, it will increasingly involve economic factors. Though in the long run this will prove true with respect to Vietnam, to date the fact that Vietnam is desperately underdeveloped and needs much foreign assistance has not caused the Hanoi Politburo to behave in an economically rational manner, nor has it significantly influenced either foreign or domestic policies.

NOTES

1. The primary source of data for this article is the Indochina
Archive, University of California (Berkeley), principally "File
7-A: Vietnam Foreign Relations/U.S.)," consisting of about
15,000 pages. Early postwar bibliographic data consulted includes Herman Kahn and Thomas Pepper, "United States
Relations with Vietnam," Hudson Institute Report, December
1976, and "Vietnam: 1976," a report to the Senate Foreign
Relations Committee by Senator George McGovern, March
1976. The best single source of material on this subject probably is the U.S. Congress: "Claims against Vietnam," House
of Representatives Report dated 30 April 1980 (outlines the
legal issues involved with United States nationals' losses in-
curred through nationalization in Vietnam and information
on Hanoi's assets frozen in the United States); "Indochina," a
report released by the Senate Foreign Relations Subcommittee on East Asia and Pacific Affairs, 21 August 1978 (contains
a 150-page study by the author entitled "Vietnam's Future
Foreign Relations," which includes a chapter on United
States–Vietnam relations); "Adjudication of Claims against
Vietnam," by the House Subcommittee on Asian and Pacific
Affairs, 27 July and 25 October 1979 (contains background
material on losses by American individuals and companies
through nationalization of property in Vietnam and on Hanoi
assets frozen in the United States); "Vietnam-U.S. Relations:
The Missing in Action and the Impasse over Cambodia" by
Robert Sutter, Library of Congress Congressional Research
Service, Issue Brief IB 87210, 2 February 1988. Recent analysis
includes "Review of U.S.-Vietnam Issues" by Assistant
Secretary of State Gaston Sigur, testimony before Congres-
sional hearing, 28 July 1988; Department of State Current
Policy Statement No. 1098, August 1988; also Frank Tatu,
"U.S. and Vietnam: Converging Interests?" *Indochina Issues* 81
(May 1988). Sources for recent information on the debate
over possible United States recognition of Vietnam include:
John Le Boutillier, "Coming to Terms with Vietnam," *New York
Times Magazine,* 1 May 1988 (a conservative argues that recog-
nition is in the best long-range United States interest);
Frederick Z. Brown, "Sending the Wrong Signal to Hanoi,"

Carnegie Endowment for International Peace, *Asian Wall Street Journal Weekly,* 18 April 1988 (a retired American diplomat argues that recognition must be part of a package deal and cannot be separated from Kampuchean peace and other issues); Walter Friedenberg, "Relations with Vietnam Backed by a Bipartisan Group on Hill," *Washington Times,* 25 May 1988 (report on the activities of six senators and four representatives, both Democrats and Republicans, who mean to force a changed Vietnam policy on the administration); Neil A. Lewis, "An Ex-P.O.W. Leads Drive for Hanoi Ties," *New York Times,* 1 June 1988 (a report on the recognition issue told through the person of Senator John McCain, conservative and ex-POW who is perhaps the most influential figure on Capitol Hill to weigh in supporting recognition); George Black, "Republican Overtures to Hanoi," *Nation,* 4 June 1988 (wrap-up on who on Capitol Hill has said what about United States diplomatic recognition of Hanoi); Nayan Chanda, "Straws in the Wind," *Far Eastern Economic Review,* 9 June 1988; Larry Pressler, "We Can't Isolate Vietnam Forever," *New York Times,* 23 May 1988; "Sending the Wrong Signal," *Bangkok Post,* 11 April 1988 (a Thai editorial warning that United States recognition of Hanoi is not simply a bilateral matter, but that Thailand and others have a vested interest; Thai Foreign Minister Sitthi is quoted as saying Thailand strongly opposes establishment of a United States interest section in Hanoi). See also the author's chapter, "American-Vietnamese Relations," in *Defense Planning for the 1990s,* ed. William Buckingham (Washington, D.C.: National Defense University Press, 1984).

2. The metaphor often used to explain this concept is the equilibrium of a chemical solution such as salt dissolved in water: the solution is in equilibrium, in balance, is relatively stable; it is a product of several elements; and while it is not permanent, it is not easily changed.

3. This was the Allied rationale against Japan in World War II, the meaning of the United Nations' acting in concert in Korea, and the basic reason why five Pacific Basin nations sent troops to fight in Vietnam.

4. This thesis is argued in Douglas Pike and Benjamin Ward,

"Losing and Winning: Korea and Vietnam as Success Stories," *Washington Quarterly* (Summer 1987).

5. In *Decision against War: Eisenhower and Dien Bien Phu* (New York: Columbia University Press, 1988) Melanie Billings-Yun argues that although in 1954, with the French losing at Dien Bien Phu, there was pressure on the United States to intervene, it was never a close call. Washington policy makers, including most right-wing hawks, recognized a lost cause when they saw one. Rather, the problem was tactical, how to avoid having the stigma of French defeat stain a Republican administration. The Eisenhower-Dulles team emerges as more Machiavellian than fits its general reputation.

6. For full treatment of *dau tranh* strategy see chapters 9, 10, and 11 in the author's *PAVN: People's Army of Vietnam* (San Francisco: Presidio Press, 1986).

7. Its position was based on the 1973 Paris agreements that ended the Vietnam War. In the summer of 1973, as part of the agreements, representatives from the U.S. Agency for International Development and North Vietnam held technical-level meetings in Paris on economic assistance to Vietnam, to which the United States had agreed as part of the "binding up of wounds of war" effort in the Paris agreements. At one meeting Hanoi presented a list of desired United States–assisted reconstruction projects, totalling about $3.25 billion. In another document, the United States acknowledged North Vietnam's economic need and implied American assistance was forthcoming. However, the American negotiators stressed they did not have the authority to commit the United States to granting $3.25 billion since this was a power reserved to the Congress; further, that the United States considered economic assistance to be dependent on Hanoi's military restraint in the South. If the Paris agreements are the touchstone for future United States–Vietnam relations, then, rather than the United States having obligation, Vietnam at the least owes the United States an apology for violating the agreements. Whatever their meaning, the agreements clearly stipulated no PAVN troop augmentation after February 1973. Yet by April 1975 virtually the entire North Vietnamese army was in South Vietnam, a total breach of the agreements.

8. For full treatment of this subject see the author's *Vietnam and the USSR: Anatomy of an Alliance,* (Boulder, CO: Westview Press, 1987).

9. As employed here, *regionalism* means supranational but geographically limited institutions and consciousness; it is both organizational and psychological.

10. The current size of PAVN is estimated to be 2,900,000, of which 1,200,000 are "full military" and 1,700,000 are "paramilitary." (Many of the latter are full-time soldiers, however.) PAVN has about 250 combat aircraft (chiefly helicopters) and about 100 small-size naval craft.

11. Western intelligence reports fix PAVN strength in the China border region at about 600,000.

12. Vietnamese assets frozen in 1975 totaled about $90 million at the time. The cash was invested and, with interest, probably totals more than double that amount today. However, the sum is about equal to the claims by United States citizens and companies (roughly half of the total by oil companies) for expropriation losses in South Vietnam.

13. The most promising commodity for trade is petroleum. Vietnam has oil reserves, although how extensive they are is not known.

14. The word "normalization" is a slippery one when applied to international relations. Consider the variation in "normal" relations the United States has with Canada, Israel, the Soviet Union, and South Africa. Properly used, "normal relations" means the establishment of an official government-to-government representation at some specific level, which can range from the lowly interest section to the fully staffed embassy.

SEVEN

United States–Philippine Security Relations and Options in a Changing Southeast Asian Context

Lawrence E. Grinter

Introduction

> No country should imagine that it is doing [the United States] a favor by remaining in alliance with us. . . . No ally can pressure us by threat of termination; we will not accept that its security is more important to us than it is to itself. . . . We assume that our friends regard their ties to us as serving their own national purposes, not as privileges to be granted or withdrawn as a means of pressure.
> —Secretary of State Henry Kissinger, June 1975

> By offering the US [use of Philippine bases], it is estimated that we have saved that country billions, tens, perhaps hundreds of billions of dollars that would have to be spent to replace the facilities with additional carrier battle groups and communications establishments.
> —Secretary of Foreign Affairs Raul Manglapus, March 1988

The controversy over burden-sharing occupies a growing place in the current discussion of United States alliances. In East Asia the debate is evident, and it complicates United States relations with our principal allies and friends in the region: Japan, South Korea, the Philippines, Thailand, Australia, and New Zealand. Each of these countries shows a different degree of willingness or reluctance to share security responsibilities and costs with the United States, as well as hesitancy to look past its own borders (and immediate ground, air, and sea space) to threats from other countries. However, only one of our Asian-Pacific allies—the Republic of the Philippines—insists on American military and economic aid in return for being defended by the United States and for hosting the American forces that protect it and underwrite the strategic balance in East Asia.

Another difference between the government of the Philippines and the governments of the other Asian-Pacific countries with which the United States has security ties is that the Philippines undertakes essentially no external defense of its own, and never has, evidently always content to rely on the United States for protection.

The economic facts of this unique-to-East-Asia "protection plus payment" arrangement are well known and are not disputed by the Philippine government: Direct United States military spending inside Subic Bay Naval Base and Clark Air Base is over $550 million a year.[1] Add in the special contracting and indirect services used by the Americans, and United States spending, whose impact on the surrounding Filipino communities is great, is probably over a billion dollars a year, or at least 3 percent of the Philippine GNP.

The review of the amended Military Base Agreement that was conducted in 1988, and that required six months of negotiations, leaks, break-offs, and resumptions in Manila and Washington, finally produced a near tripling of United States–pledged future payments to the Philippines for use of the bases: up from $180 million to $481 million per year.[2] The new arrangement was signed on October 17, 1988, and runs until September 16, 1991, when the original 1947 agreement, and a related 1979 amendment upon which it is based, ends. Circumstances surrounding the 1988 review also produced a willingness by the United States government to seriously consider taking our military out of the Philippines. Secretary of State George Shultz warned Manila that it could price itself out of the market, and at the signing Shultz told Philippine Foreign Secretary Raul Manglapus that both sides were now "keeping their options open."

Going into the base negotiations, Secretary Manglapus had given a speech at a Manila economic forum in March 1988 in which he drew a stark contrast between an "American colossus" that used other countries' territory for bases, and the Philippines, which stood "alone" in Southeast Asia, accommodating United States desires. Manglapus' conclusion: the Philippine bases were obviously worth "tens, perhaps hundreds of billions of dollars" to the United States, and Washington should expect to pay much more for Manila's "accommodation" to American "global strategy."[3] That is how the government of the Philippines, or at least its foreign minister (President Aquino's involvement in national security and foreign policy matters is not believed to be strong), characterized a negotiation between two allies who for almost a century have been closely associated in peace and war, including a brave joint effort against Japanese imperialism. The military bases issue thus clarifies how the two countries' governments view their security priorities and responsibilities in the context of a changing Southeast Asia. This essay explores this relationship and its difficulties and then suggests options for United States policy.

The Evolving "Special Relationship"

The United States acquired the Philippines and Guam by the Treaty of Paris on December 10, 1898, which ended the short-lived and highly nationalistic Spanish-American War. The subsequent United States–Philippine relationship conformed to no preexisting pattern of colonial administration; it was a special arrangement involving what the United States Supreme Court termed "unincorporated territory."[4] With American military power in the Western Pacific backing up the new vision of Manifest Destiny, American authorities helped by conservative Philippine elites (called "collaborators" by their rivals), quelled the Aquinaldo insurgency—and put an end to the short-lived Malolos Republic—with great dislocation and loss of Filipino life.[5] The two countries then entered upon their colonial dependency relationship, which has been described as a kind of "compadre colonialism," justifying imperialism as altruism and oligarchical conservatism as evolutionary nationalism.[6] From the start, Washington and most of the American public viewed their administration of the Philippine Islands as a temporary tutelage to prepare Filipinos for independence. Nevertheless, American ambivalence about colonial administration, contradictions and fragmentation in Philippine politics, and the evident benefits of the relationship to both sides delayed independence. The intense nationalism expressed in the Bonifacio-Aquinaldo insurrection was stifled and its natural outcome postponed for many years. The Japanese invasion further delayed independence, and the war itself caused terrible damage to Manila and other urban areas.

Despite its violent beginnings, American patronage of the Philippines left a heritage that included the linking of the Philippine and American economies and the Philippines' complete reliance on the United States for external security. The United States promoted an American-style political system (despite its lack of roots in Filipino tradition), the use of English as the common language, and complete acceptance of the goal of universal education, which greatly reduced illiteracy. Less positive were the imposition of an American social overlay on the Hispanicized and heterogeneous Philippine culture, and the perpetuation in power of the oligarchical land-owning elite.[7] Like other unequal relationships, the United States–Philippine association had its "love-hate" aspects, accentuated in this case by fragmentation of the Filipino political elite into conservative pro-American groups on one side and radical nationalistic groups in opposition. Drawn by cultural and economic elements of the American influence, many Filipinos nevertheless have been irritated by the sense of superiority and the arrogance that some Americans convey toward them. American enjoyment of Filipino culture and satisfaction with their overall tutelage sometimes contrasts with their disappointment that Filipinos somehow do not "measure up" to American expectations. Filipinos reciprocate with criticism of American troop behavior and expressions of fatigue at being the only ASEAN country to host large foreign military forces. Basically, the Filipinos have wanted respect, and the Americans have wanted appreciation. Neither side has gotten enough of what it sought.[8]

Between World Wars I and II, Filipinos (with American guidance) drafted their own constitution, remodeled their legislature, instituted military training, and built armed forces with General Douglas MacArthur as principal military adviser (and de facto chief of staff) to the charismatic, and unpredictable, veteran nationalist commonwealth president, Manuel Quezon. But fear of rising Japanese militarism persuaded Philippine authorities to seek a permanent, dominion-like relationship with the United

States. A few hours after attacking Pearl Harbor on December 7, 1941, Japan also invaded Hong Kong, Singapore, and the Philippines. On May 6, 1942, with the capitulation of the last American and Filipino regular forces on Corregidor, the country came under Japanese control. Some Filipinos collaborated with the Japanese. Others took to the jungles and fought as guerrillas against the brutal Japanese occupation. When General MacArthur's forces began the liberation of the islands at Leyte, the Philippine government-in-exile returned, and the battle with the Japanese was joined. Manila was ruined, and much of the country's urban infrastructure and dwellings were destroyed. It took many years for the Philippines to recover from the war, and American assistance in rebuilding Japan was resented by some Filipinos.[9] In May 1946, Manuel Roxas was elected the first president of the new Republic of the Philippines, and the country became formally independent on the Fourth of July, 1946.

Despite the colonial relationship and the inevitable strains through the years between the two countries, there has never been a break of diplomatic relations, and numerous ties and unique associations have continued to bind together the United States and the Philippines. These include:

- Benefits to Filipino veterans for service in the United States armed forces (exceeding $100 million in Fiscal Year 1988)
- United States social security payments to Philippine citizens
- Large numbers of Filipinos with established or latent claims to American citizenship (Over 1.7 million ethnic Filipinos reside in the United States. Numerous members of the Marcos and Aquino governments own homes and other property in the United States.)
- Many United States educational programs in the Philippines, and the continuing attraction of educational and training opportunities in the United States
- Considerable Filipino travel and immigration to the United States
- A variety of complex business and financial arrangements beneficial to both sides (United States private investment in the Philippines is in the billions, and the private resident American community of approximately 25,000 is the largest in any Asian country.)[10]

Every postwar Philippine president has talked about ending the "special relationship" with Washington—it being characterized by Manila authorities as unequal and patronizing. Ferdinand Marcos at times was shrill in this regard, though his actions never matched his rhetoric; the economic benefits of the bases and other aspects of the American largess were too enticing to Marcos and his friends—as later became tragically evident—to risk a formal break. But there exists a continuing tension between Filipinos and Americans on the bases issue. It derives from the historic encounter in the late nineteenth century between a growing American nation destined to play a major power role in the Pacific and the Philippines, a small country wrestling with its own national identity and purpose.

Manila and Washington: Divergent Security Agendas

Of all aspects of the Manila-Washington relationship, the military bases controversy clarifies best the different security perceptions and agendas of the United States and the Philippines. American administrations have primarily viewed bases in the Philippines as convenient pieces of real estate from which United States naval and air power can be projected into the South China Sea and Indian Ocean on behalf of the United States and its friends. As William Sullivan, a former American ambassador to the Philippines, writes,

> They are first of all part of a strategic logistics line that reaches through Diego Garcia into the Arabian Sea. They are also in excellent position to service the United States fleet as a counterpoise to Soviet surface and submarine activity in the China Seas. . . . In this sense, the whole Pacific community is a beneficiary of the current Philippine-American strategic relationship.[11]

But for many of the Filipino elite, who represent a country with no external security responsibilities, the bases are seen as territory made available for another country's use at both inconvenience and some risk. In 1986, the Philippine ambassador to the United

States, Emmanuel Pelaez, contrasted Philippine security perceptions with those of the United States, which he said stations three-quarters of a million troops overseas in forty different countries.

> The Philippines is a developing country whose GNP of around $40 billion is but 1.3 percent of the United States GNP. The Philippines needs to spend at least $2 billion annually, or about one-half of its export receipts, to pay interest on a $26.3 billion foreign debt. About two-thirds of Filipinos now live below the poverty line as a result of falling real incomes under the previous regime. . . . In short, the Philippines faces an agenda of development and national-interest concerns basically different from the priorities the United States faces as an immense economic-military power.[12]

Moreover, while Filipinos take comfort in American protection (if not the presence that produces that protection), they have no tradition of strategic military planning. Some even continue to rationalize Japan's 1941 attack as resulting from the Philippines' being a United States protectorate rather than because of the country's geopolitical location and importance in Japan's strategy. Finally, there is anger and confusion about the Marcos legacy, which robbed the Philippines of much self-respect and sums of money possibly equivalent to half or more of its GNP.[13] Thus the Aquino government, administering a recovering democracy reflecting all these crosscurrents and contradictions, simply has not been able to come to terms with the security requirements of surviving independently in the changing Southeast Asian security context. President Aquino's government is spending about 9 percent of its annual budget on defense; that is what remains after social programs are funded and another 45 percent of the budget goes to interest payments on the gigantic foreign debt Aquino inherited from Marcos.[14] With this background, when Mrs. Aquino commented on national defense to a group of ASEAN delegates in a December 1987 meeting, she said, "By virtue of geography alone the Philippines can be indifferent to external threats."[15] A comfortable perception, no doubt. Undoubtedly Filipinos expressed similar views in 1939 and 1940.

Given the undeniable need for economic recovery and reform

in the Philippines, it is nevertheless startling to Americans to encounter statements such as President Aquino's or to glimpse how complacent, or uninformed, many Filipino leaders are about security developments near their country's borders. It is difficult for most Filipinos to look beyond their fixation with Marcos and the United States to their own region, the South China Sea and Southeast Asia. For a country so long associated with a superpower who was also the world's policeman, for a nation that has been militarily invaded or liberated four times (by Spain, the United States, Japan, and the United States, respectively), Filipinos have one of the most myopic and complacent attitudes toward external security of any people in Asia. This, of course, is not true of all Filipino leaders. Secretary of National Defense Fidel Ramos, a West Point graduate and previous armed forces chief of staff under both Marcos and Aquino, is frank in his praise of the United States–Philippine alliance, the bases' utility, and the close relationship between the two countries' armed services.[16] Foreign Secretary Raul Manglapus, Manila's point man in the 1988 attempt to obtain new billions from the United States, has been one of the few Filipino leaders to explore the concept of a regional burden-sharing arrangement for the American military presence in Southeast Asia.[17] And the Philippine ambassador to the United States, Emmanuel Pelaez, in writing of the Japanese conquest and the resulting devastation, speaks of the "terrifying lesson in the importance of strong external defense."[18]

But these realistic views do not translate into realistic military budgets, or hardware, or new ships or planes or decent salaries for the soldiers fighting the guerrillas. The headquarters building for the Armed Forces of the Philippines (AFP), at Camp Aquinaldo, lies in burnt-out ruins from Colonel Honasan's August 1987 coup attempt and the additional damage inflicted in the December 1989 mutiny. By contrast there are gleaming skyscrapers in the Makati business district and even private residences in the Forbes Park suburb that are worth more than it would take to rebuild AFP headquarters. And so the myopia is perpetuated and the post-Marcos fixation continues despite the changing and not entirely benign situation in Southeast Asia. Consider, for example, these facts:

- In spite of Gorbachev's rhetoric, the Soviets patrol the South China Sea and violate Philippine air and sea space.

- Vietnamese troops have repeatedly crossed the Thai border.
- The United States has severe budget constraints, and its military assets in the Asian/Pacific/Indian Ocean theaters are stretched from Yokosuka to the Arabian Gulf.
- All ASEAN countries except the Philippines have converted their defense doctrine and acquisition efforts from a focus on counterinsurgency to one on external defense, and some are spending very large proportions of their GNP on defense (Brunei and Malaysia allocated about 15 percent and 7 percent, respectively, in 1987).
- The Philippines has no capability to protect the sealanes of communication (SLOCs) or oil tankers upon which its economy depends.
- The Philippines has a serious territorial problem with an immediate neighbor, the much better armed Malaysia.
- Armed scrambles for territory and oil shelf rights periodically erupt in the South China Sea involving Vietnamese, Chinese, Indonesian, Malay, and some Philippine facilities and troops.

Taken singly, any of these factors might be dismissed as not particularly threatening to the Philippines. Examined in combination, no state in the region can afford to ignore them or draw the wrong conclusions from them. Those who would ignore them, like Senator Leticia Ramos Shahani, the chairperson of the Philippine Senate Foreign Affairs Committee, simply further perpetuate the myopia.[19] In these leaders' views, the bases are simply money-earning devices that, even given that benefit, still have to be phased out so Filipinos can regain their self-respect.

The Countdown to 1991–1992: Aggravating Factors

On September 16, 1991, the current amended, recently revised, Military Bases Agreement (MBA) between Washington and Manila will terminate.[20] In the summer of 1992 a new Philippine president will be elected. In the nine months between those events it is anybody's guess what might emerge: a new treaty covering the bases, a simple extension of the current agreement with no new treaty, or withdrawal of United States forces and turnover of

American facilities to the Government of the Philippines. Ultimately it will depend on the policies and makeup of the Philippine government in office in 1991 and how the United States deals with the problem. Assuming Mrs. Aquino remains in office and is not overthrown by the armed forces in the meantime, a variety of developments bearing on the United States–Philippine security relationship and the basing problem will come up. Let us briefly survey them.

Philippine Domestic Politics

During 1988 and 1989 there was much preliminary maneuvering in Manila to establish the major contenders for the 1992 election, after which President Aquino must step down in accordance with the new constitution's prohibition on more than one six-year term for the president.[21] The two most prominent early associates of Mrs. Aquino, ex-Defense Minister (now Senate minority leader) Juan Ponce Enrile and current Vice-President Salvador Laurel, broke with her in 1987 and 1988 and are joined in an anti-Aquino tactical alliance. Other potential presidential candidates in the Senate and House include Senate President Jovito Salonga, leader of the Liberal party, House Speaker Ramon Mitra, and Senator Aquilino Pimentel, chairman of PDP-Laban. Defense Secretary Fidel Ramos is also expected to run. The attacks on Mrs. Aquino and her ambivalent leadership have seen Philippine politics revert to the personality-driven and often irresponsible invective of the past. Whether the issue is Mrs. Aquino's leadership, her family's wealth, government scandals, the huge foreign debt, the ineffective land reform efforts, the poor pay and conditions in the armed forces, or the future of the bases and the United States relationship, the opposition dominates the political debate. With twenty-seven daily newspapers in Manila constantly pouring out the political acrimony, the opposition's negativism has set the tone for the 1991 run-up to the election and for termination of the Military Bases Agreement. In Fred Greene's words,

> Because of economic strains, the stress of Philippine politics, and the unavoidable diminution of [her] influence as the 1992 date for the end of her term approaches, Mrs. Aquino's dominant position is likely to erode. As noted, her "party" encompasses a wide spectrum of views, and

many of her supporters oppose a new bases arrangement. With an unwieldy coalition that is difficult to manage, decision-making becomes slow and uncertain, a condition that will intensify as others jockey to succeed her.[22]

Leadership and legitimacy, then, are the most critical, and least predictable, factors bearing on the transition period of 1991–1992.

The Economy

If the trends prior to the December 1989 coup attempt resume (a 6 to 7 percent GNP growth rate and new foreign investment of about $800 million per year) and the Philippine government can reschedule and consolidate its foreign debts, macroeconomic issues should not unduly influence the 1991–1992 debates. The Aquino government has gained a near tripling of pledged American military and economic assistance (from $180 million to $481 million per year), which began in October 1989. The World Bank, the IMF, and the Asian Development Bank, as well as Japanese and American private banks, are supportive of Aquino's goals.[23] The Central Bank governor and the finance minister have been successful in debt rescheduling so far, and the more radical views of officials like Solita Monsod and Leticia Shahani, who have argued for debt repudiation, are not government policy. Foreign investment, particularly intra-Asian investment, is a key to the Philippines' economic recovery; regional assistance to the Philippines, especially from the ASEAN countries, ought to be growing. (American investors are more inhibited by political risk and the bases controversy than are Asian investors.)

But it is the land reform problem and the vested interests' opposition to it that has produced the greatest economic failure, so far, of the Aquino years. The president chose not to take a strong position on land reform, allowing the initiative to pass to the legislature, where it has been debated and deadlocked—not unusual in the Philippines—for many months. The oligarchical interests in the lower house, coupled with corruption and inefficiency in the countryside, have thus far prevented any major redistribution of land to individuals.[24] The land reform problem also joins other serious structural inequities in the Philippine socioeconomic system: income distortions; a fractured, often irresponsible political and business elite; inefficient industries;

critical ethnic/religious conflicts; a violent culture in which two serious insurgencies float; underpaid and mutinous armed forces; a foreign debt equal to 80 percent of the GNP; no oil or gas resources; and a long-term population-increase problem about which nothing significant is being done.[25] The Philippine population is now at 60 million and is likely to be pushing 85 million at the close of this century.

Still, economic progress has been made; if the country's industries and productivity can get back on their feet, and if the armed forces can be held in check, there is a chance that personal incomes can be generated at levels that will begin to compensate for population growth and foreign debt. It is a race against time.

The Insurgency

The communist insurgency in the Philippines, while losing momentum in 1988 and 1989, did not collapse, and the endurance of the 35,000 party members and 24,000 guerrillas[26] is sure to keep the rebellion alive well into the 1990s. Inheriting the insurrection from the Marcos regime, Mrs. Aquino applied a variety of carrot-and-stick measures against the communists, none producing much success, while her government was shaken by political and military factionalism amid the wider tremors in the Philippine society and economy. The late 1986-early 1987 two-month government cease-fire with the communists failed to weaken the rebels and, indeed, gave them a public forum in which to demonstrate their appeal and personality. Honasan's August 1987 coup attempt paralyzed further government security operations, and it was not until the spring of 1988 that Manila could reapply pressure on the guerrillas.

In February, March, and July 1988, important communist leaders and documents were captured; finances and communist front organizations also were exposed.[27] Overall Philippine armed forces (AFP) strategy and tactics against the Philippine Communist party (CPP)/New People's Army (NPA), however, were slow to change, and in 1987 and 1988 the Aquino government found itself openly relying on vigilantes to provide local security where government forces and political control could not reach. The captures of CPP/NPA leaders in 1988 had a positive, if temporary, effect on armed forces and police morale; they brought relief to the security situation in Manila, which had been reeling under the NPA "sparrow squad" urban siege. However, one very prominent detainee, Romulo Kintanar, head of the NPA, escaped from custody in November 1988. Kintanar may have directed the urban hit squad that killed the United States military adviser, Colonel James Rowe, in Manila on April 21, 1989.[28]

In August 1988, the Philippine Defense Ministry published a new doctrine and organization plan for combating communist forces entitled "The Three-Tiered Defense System for Internal Security." The new system links regular AFP brigades, battalions, and special action forces ("Military Mobile Forces") with the constabulary, police, and local citizen militias now called Citizen Armed Force Geographical Units, CAFGU ("Territorial Forces") and local militia and interest groups ("Civilian Volunteer Organizations," CVO).[29] Vigilantes are not mentioned, but clearly the forming of the new CVOs and CAFGUs is an attempt to contain the vigilante phenomenon, which mushroomed after Marcos's overthrow. Nevertheless, it is believed that a majority of the vigilantes, and certainly the landlords' private militias, continue to operate independently.[30]

How the three echelons of the AFP defense system actually operate in various localities under communist pressure depends, of course, on the nature of the threat as well as on local and regional politics. The new defense system seems well thought out. It is also clearly an accommodation to the existing problems and exigencies of the Philippines' internal security situation. The new strategy acknowledges that in the past the various units and activities have operated independently of each other, negating much of one another's effectiveness. Without an ability to link local citizen defense groups to the territorial and mobile forces, duplication of effort and serious gaps in protection occurred. Today, Manila is trying to link up all three defense echelons by appointing "peace and order councils" to coordinate the activities. The aim is to put the resources of the 190,000 soldiers and police in the Philippines, and the 9 percent of national budget (or 2.1 percent of GNP)[31] that goes for defense, at the disposal of national authority.

In summary, until the December 1989 mutiny, armed forces revitalization and doctrinal streamlining coupled with CPP/NPA losses had bought time—valuable time, to be sure, but time only—until the more serious distortions and grievances in the Philippines are tackled.

Recurring Bases Issues

With the Military Bases Agreement terminating just as the next Philippine presidential election campaign begins in the fall of 1991, it is certain that the acrimony and debate of the campaign will distort the bases issue; the positive dimensions of the American presence and the defense tie may be drowned out in the rhetoric on the bases. While a majority of Filipinos will probably continue to support retaining the American bases,[32] the politicians distancing themselves from Aquino will use the bases as a whipping boy in the likely-to-be-incendiary political campaign.

The bases debate is also certain to cause the re-emergence of those issues in the 1988 review that were unresolved or left "for further consultation," such as criminal jurisdiction and Philippine influence on military operations at the bases. The nuclear weapons issue also is sure to be revived, and leading opponents of nuclear weapons and the bases, like Philippine Senator Wigberto Tanada, have lost no time demanding clarification of the nuclear provisions in the new agreement and pushing for nuclear-free zone legislation.[33]

Regarding nuclear weapons, the 1988 agreement stated that "the storage or installation of nuclear or nonconventional weapons or their components in Philippine territory shall be subject to the agreement of the Government of the Philippines." It also stated that "transits, overflights or visits by US aircraft or ships in Philippine territory shall *not* be considered storage or installation."[34] This language is conducive to American interests and represents a clear accommodation by the Aquino government to current United States nuclear policy.

Armed Forces Loyalty

In early December 1989, an eight-day coup attempt almost toppled the Aquino government. The judicious use of United States airpower, requested by the Manila government at the extreme moment, was instrumental in rescuing Aquino from this near-fatal mutiny, the sixth and most serious against her in four years in power. At least 3,000 troops participated in the uprising, including the elite Scout Rangers. It was a very close call—the presidential offices, armed forces headquarters, and the financial district were all attacked. The event vividly showed the weaknesses in the Philippines: continuing deep divisions in the armed forces, weak leadership by President Aquino and Secretary Ramos, and an economy that got stopped dead in its tracks. In Senator Richard Lugar's words, "Without any doubt the coup . . . . is a disaster for economic development."[35] Very few of the rebels were captured or killed, most simply slipping away in civilian clothes or returning to their barracks amid cheers.[36]

If there was one bright spot in December's dramatic events, it was the Bush administration's unwavering support for the Aquino government in its most difficult moment. Whatever the nationalist reverberations among the opposition in Manila, the Bush action gained real appreciation from the Aquino government, which is expected to factor into the upcoming bases renegotiations. But the coup also showed that the greatest immediate danger to democracy in the Philippines and to the stability of United States–Philippine security relations is not the communist threat, or the economy, for that matter, but the wavering loyalties of the Philippine armed forces.

Whither United States Policy?

Clearly the United States enjoys a better and more relaxed relationship with the Aquino government than it did with Ferdinand Marcos. The American ambassador is evidently no longer wiretapped or shadowed, nor is he prevented access to the Philippine president. Marcos, whose primary motive now appears to have been the exploitation of the oligarchical system from which his personal wealth and that of this family and friends grew, kept United States policy on a tight rein. The trade-off was simple and usually effective: Washington should not get too critical of Marcos and his corruption, or continued American access to the bases would be jeopardized. Dealing with President Aquino and her reformist government is, compared to that, a breath of fresh air. Mrs. Aquino is honest and moral (if some officials around her are not), and she represents the forces of democratic reform in both government and society.

However, the new compatibility between Manila and Washington has eliminated neither the Philippines' major problems nor the differences between Washington and Manila on issues like the counterinsurgency, land reform, and the bases. As the 1991–

1992 period approaches, the Bush administration will have to make choices about how to deal with the Philippine government on a variety of issues including, within the security context, the bases. Let us examine the options.

Extending the Current Agreement

With or without a new treaty, the most likely American policy will be to try to extend the October 1988 agreement past the September 1991 termination point. There does not have to be a new defense treaty, even though the Philippines' constitution calls for it. It can be put off or held in abeyance. Under this scenario, there would be some kind of extension of the current 1989–1991 financial and operational arrangements. The Filipinos would undoubtedly press for more compensation, the Americans would try to hold the line at about $500 million, and both sides would fine-tune arrangements on status of forces, criminal jurisdiction, Filipino access to United States military operations out of Subic Bay and Clark, and the nuclear issue. As the 1988 negotiations were nearing agreement, United States Ambassador to the Philippines Nicholas Platt sought to prompt interest in this kind of future situation during the interviews and meetings in Manila.[37] Platt will head the new renegotiation on the American side.

As part of its future negotiating strategy, the United States will have a variety of arguments and options it can call upon to influence the Filipino position. Drawing the line at $500 million per year is one. The "neither confirm nor deny" nuclear policy is another. Other measures are available that could be aimed at mollifying Filipinos on questions of criminal jurisdiction, social issues, labor agreements, and squatters' rights. However, there are three external considerations that the United States needs to examine; heretofore, they have been largely ignored. They involve the ASEAN governments and the Soviet Union. Active pursuit of these out-of-country options and arrangements brings the bases problem into a Southeast Asian regional context. The effect of emphasis on the wider context would not be lost on Manila.

ASEANizing the Bases

United States policy makers ought to start a serious dialogue with the other ASEAN governments about a *multilateral* military presence at Subic Bay and Clark. If Raul Manglapus' comments are valid about the Philippines' feeling "alone" in hosting the American military presence, then Washington ought to open its facilities in the Philippines to more naval visits and air force exercises by other ASEAN armed forces. This is already being done at Clark as part of the Cope Thunder tactical air exercises, which regularly involve the Thai and Singapore air forces. In turn, ASEAN ships could be encouraged to regularly visit Subic Bay. This would reduce the visibility of the United States presence and add a *regional* military character to Subic Bay and Clark. And it would contribute to the regional "political burden-sharing" notion that the Philippine foreign minister has been talking about.

Limited United States Redeployment

Another and not unrelated consideration to bringing more ASEAN armed forces into Subic Bay and Clark is to begin a limited American military redeployment *out* of Subic Bay and Clark to other ASEAN installations. Good military facilities exist in ASEAN: In Thailand, Utaphao and Sattahip are excellent military bases and are clearly underutilized. In Singapore, the large Sembawang harbor in the Johor Straits could host some U.S. Navy assets. Possibly Brunei would accept a rotation of U.S. Seventh Air Force units. If the Philippine government cannot cope with being the "lone" ASEAN government to host American forces (in spite of having a bilateral defense treaty with the United States and receiving over a billion dollars per year in United States military spending), I suspect there are other ASEAN governments that would not refuse such benefits. Singapore has already signalled interest in a limited United States redeployment to Singapore.[38] Moreover, this option or an alternative policy track would spread the security responsibility of hosting American defense forces to other ASEAN governments.

Gaining an Understanding with Moscow

On two occasions, Mikhail Gorbachev has proposed a trade-off of United States and Soviet military facilities in the Philippines and Vietnam, respectively. His most explicit offer was at Krasnoyarsk, in September 1988, in one brief sentence on the issue. His proposal was immediately seized upon by the Western media, particularly the *New York Times* and the *Washington Post*, as essentially a complete and symmetrical trade-off of the Soviet military presence in Vietnam for the United States military presence in the Philippines. In fact, Gorbachev proposed nothing of the kind. What he actually said was this:

> If the United States eliminates its military bases in the Philippines, the USSR will be prepared, in agreement with the Government of Vietnam, to give up the naval material and technical support point at Cam Ranh Bay.[39]

Note the *asymmetry* in this proposal: The United States is to completely vacate all Philippine military installations it uses in return for the Soviets' giving up a *support point* at Cam Ranh Bay. What about Soviet air force installations at Tan Son Nhut and Da Nang? What about the huge Soviet intelligence collection station in Vietnam? What about Soviet advisers there and military assistance? And, given the impending termination of the MBA between Washington and Manila, what about the twenty-five-year defense treaty (with fifteen years still to go) between Moscow and Hanoi?

In short, Gorbachev was testing public opinion and, given initial reactions in Manila and the United States, he did pretty well. In fact, the United States would be foolish to vacate military facilities in Southeast Asia without a complete and permanent end to the Soviet presence in the area and termination of other threats to the SLOCs, the flow of oil, and the commerce from the Indian Ocean and the Persian Gulf upon which East Asia depends. Bases at Subic Bay and Clark are critical links in the United States logistics line that stretches from San Diego and Pearl Harbor through the Philippines to Diego Garcia and the Persian Gulf. Gorbachev's proposal would break that line at its center in exchange for some kind of partial Soviet military disengagement from Vietnam's coast. Lack of interest in the proposal by both the Reagan administration and the ASEAN governments was not, evidently, lost on Moscow: In a later, December 1988, visit to Manila following his trip to Japan, Eduard Shevardnadze told Raul Manglapus that the Soviets recognized the United States had long-standing political and economic interests in the Philippines and that Moscow might, at some future point, consider a unilateral reduction of forces in Vietnam.[40]

There are, of course, other reasonable tension-reduction measures involving the Philippines that the United States could suggest to Gorbachev as part of a de-escalation package and a testing of Soviet sincerity. First, Mr. Gorbachev could end Soviet and Soviet client support for the NPA in the Philippines. Second, he could stop trying to get Soviet intelligence ships, disguised as merchant vessels, inside Subic Bay and into other Philippine waters.[41] Third, the Soviets could stop efforts to get dry dock facilities, with related electronic monitoring capabilities, built near Manila and Subic Bay. Fourth, Soviet aircraft could refrain from violating Philippine air space. Fifth, the Soviets could end plans to increase their KGB and intelligence collection operations in their embassy in Manila. Any of these changes would show Soviet good faith. Once these pressure tactics ended, the United States and the Soviet Union could then enter into broader arms control, deescalation, and nonintervention negotiations involving the South China Sea area.[42]

In conclusion, the 1991–1992 renegotiation of the United States–Philippine military bases agreement will take place in a changing political context in the Philippines and a changing security context in Southeast Asia. The United States ought to shift the renegotiation from having a strictly bilateral United States–Philippine focus to a multilateral Southeast Asian framework in which more realistic assessments of interests and responsibilities are presented.

NOTES

1. United States Information Service, "Background on the Bases," 2d ed., Manila, 1988, p. 16.
2. Text of the 17 October 1988 Memorandum of Agreement between the Philippines and the United States. See *Manila Bulletin*, 19 October 1988.
3. Raul S. Manglapus, "Accommodating the US Bases: 1898–1991," a speech delivered before the Philippine Council for Foreign Relations in Manila on 28 March 1988.
4. Following Senate ratification, the Philippines became an American possession on 6 February 1899. Details are in Salvador P. Lopez, "The Colonial Relationship," in Frank H. Golay, ed., *The United States and the Philippines*, American Assembly, Columbia University (Englewood Cliffs, NJ: Prentice-Hall, 1966), pp. 7–31.
5. I estimate that about 4,200 United States soldiers and 100,000 Filipinos died (the latter mostly due to diseases and famine) in the struggle. Estimates by scholars ranging from 36,000 to 200,000 native deaths can be found, however. See, for example, D. J. M. Tate, *The Making of Modern South East Asia*, vol. 1 (New York: Oxford University Press, 1971), p. 375, and Leon Wolff, *Little Brown Brother* (Garden City, NY: Doubleday and Co., 1961), p. 11. Also see Stanley Karnow, *In Our Image: America's Empire in the Philippines* (New York: Random House, 1989), pp. 155–60, 178–80, 188–95.
6. David Joel Steinberg, *The Philippines: A Singular and a Plural Place* (Boulder, CO: Westview Press, 1982), p. 48.
7. Karnow, *In Our Image*, pp. 229–31.
8. See, in particular, Karnow, *In Our Image*, chapter 8, "America Exports Itself," pp. 196–226.
9. A familiar argument, often repeated for Americans, goes like this: "It's really fair that we get this kind of treatment. After all, after World War II, we were a devastated country. The common feeling among veterans who fought alongside the Americans is that we didn't get a square deal, that enemies like Japan got a better deal from the Americans." A ranking Filipino, as cited in *U.S. News and World Report*, 29 August 1977, pp. 29–30.
10. For background see Lawrence E. Grinter, *The Philippine Bases:*

Continuing Utility in a Changing Strategic Context (Washington, D.C.: National Defense University Press, 1980), pp. 44–45.

11. Sullivan, quoted in Carl H. Lande, ed., *Rebuilding a Nation: Philippine Challenges and American Policy* (Washington, D.C.: Washington Institute Press, 1987), p. 544.

12. Emmanuel M. Pelaez, "The Philippines and the United States," in Lande, *Rebuilding a Nation*, pp. 47–48.

13. "Dollar salting," taking hard currency and convertible securities out of the Philippines and stashing them abroad, became a major, possibly *the* major financial/economic activity of the Marcos regime in its last ten years in office. Using vast networks of cronies, agents, family members, dummy corporations, and surrogate ownership arrangements, Marcos and his oligarchy may have accumulated hidden wealth equivalent to the Philippines' foreign debt, or about $28 billion to $30 billion. See Lewis M. Simons, *Worth Dying For* (New York: William Morrow & Co., 1987), pp. 171–84.

14. Author's interviews in Manila on 8 and 9 September 1988 with National Defense Secretary Fidel Ramos and Central Bank Governor Jose Fernandez.

15. President Aquino, as cited by Secretary Manglapus in his "Accommodating the US Bases: 1898–1991."

16. Author's interview with Secretary Ramos, 9 September 1988.

17. See Secretary Manglapus' speech of 28 March 1988.

18. Pelaez, in Lande, *Rebuilding a Nation*, p. 46.

19. As she remarked to me on 8 September 1988, in Manila, when these external security realities were pointed out to her, "We don't have the money for those things."

20. If, on 16 September 1991, either Manila or Washington *gives notice* to terminate the American basing presence, the United States must be out on 16 September 1992. If neither government gives notice, the basing arrangements continue indefinitely pending future adjustments.

21. There is, however, debate as to whether Mrs. Aquino is bound by that particular constitutional restriction, since, while the constitution under which she currently serves stipulates a six-year term ending on June 30, it did not go into effect until October 15, 1986. Others point out that Mrs. Aquino actually took the presidential oath of office on February 26, 1986.

22. Fred Greene, ed., *The Philippine Bases: Negotiating the Future* (New York: Council on Foreign Relations, 1988), p. 14.

23. William H. Overholt, "Pressures and Policies: Prospects for Cory Aquino's Philippines," in Lande, *Rebuilding a Nation*, pp. 105–106, and Clayton Jones, "Philippine Economy on a Roll, but Danger Signs Ahead," *Christian Science Monitor*, 27 April 1989, pp. 1–2. President Aquino's November 9–12, 1989, visit to the United States evidently resulted in some new prospects for trade and investment but no explicit promises for increased American aid. Nor did any negotiations on the bases materialize. See *Philippine Star*, 16 November 1989, p. 1, and *Philippine Daily Globe*, 15 November 1989, p. 7.

24. See Matt Miller, "Corruption Is Bringing Manila's Land Reform Program to a Near Halt," *Asian Wall Street Journal Weekly*, 10 July 1989, pp. 1, 26; and Clayton Jones, "Mud Sticks to Aquino Government," *Christian Science Monitor*, 6 July 1989, p. 3.

25. Lawrence E. Grinter, "Taiwan and the Philippines Tell a Tale of Two Nations," *Montgomery Advertiser/Journal*, 2 October 1988, p. 3B.

26. Richard D. Fisher, Jr., "Confronting the Mounting Threat to Philippine Democracy," Asian Studies Center Backgrounder no. 67 (Washington, D.C.: Heritage Foundation, 3 September 1987), p. 3, and *Christian Science Monitor*, 1 March 1989, p. 3.

27. Among those arrested were Rafael Baylosis, current CPP secretary-general, and Romulo Kintanar, current NPA head. Also see, for an assessment, David F. Lambertson, "Future Prospects for the Philippines," Current Policy no. 1157, U.S. Department of State, Washington, D.C., 7 March 1989.

28. In early April 1989, Kintanar had annnounced that NPA units would soon launch "complicated special actions." *Far Eastern Economic Review*, 27 April 1989. Also see *Christian Science Monitor*, 3 May 1989, p. 6, and the *Manila Journal*, 11 May 1989, p. 1.

29. See "Memorandum from Secretary of National Defense Fidel V. Ramos to Armed Forces of the Philippines Chief of Staff and Chief, Philippine Constabulary/Director General, Integrated National Police," dated 11 August 1988, and the 8 June 1988 implementing instructions to Executive Order No. 264.

30. By mid-1988, vigilante groups across the Philippines may have numbered 200, with shifting, sometimes overlapping, mem-

berships possibly aggregating 150,000 men. This makes them larger than both of the ideological insurgent forces, the NPA and the Muslims, as well as almost equal to the entire Philippine armed forces and police combined! In first introducing the three-tiered defense system to the Philippine public on May 12, 1988, Secretary of National Defense Ramos emphasized that CVO "is the official term," and "not vigilantes" and "not Alsa Masa" or Nakasaka "or anything else." Justus M. Van der Kroef, "The Philippine Vigilantes: Devotion and Disarray," *Contemporary Southeast Asia*, September 1988, pp. 174, 180; and Justus M. Van der Kroef, "Organizing the Philippine Counterinsurgency: The Problem of the Vigilantes and 'Citizen's Armies,'" *Issues and Studies*, vol. 25, no. 4 (April 1989), pp. 124–28.

31. See addresses by Secretary of National Defense Fidel V. Ramos before the Asia Society and the Council on Foreign Relations, 19 May 1988, New York, mimeographed, pp. 7 and 6, respectively.

32. The Social Weather Stations polling organization in Manila indicated throughout 1987 and 1988 that only about 11 percent of the people in the Philippines wanted an immediate termination of the basing arrangement, regardless of compensation levels or other adjustments.

33. Richard D. Fisher, Jr., "A Strategy for Keeping the US Bases in the Philippines," Asian Studies Center Backgrounder no. 78 (Washington, D.C.: Heritage Foundation, 20 May 1988), p. 7. Also see *Philippine Daily Inquirer*, 7 November 1989, p. 1, and 13 November 1989, p. 1.

34. Text of the 17 October 1988 Memorandum of Agreement between the Philippines and the United States as reprinted in the *Manila Bulletin*, 19 October 1988.

35. Lugar, as cited in *Christian Science Monitor*, 8 December 1989, p. 7.

36. The "Reform the Armed Forces Movement" (RAM) inside the Philippine armed forces was largely responsibile for the rebellion against Marcos and Ver in February 1986, and, as led by dissident Colonel Honasan, against Aquino and Ramos in August 1987. Honasan loyalists and others appear to have mounted the December 1989 attack. RAM-type disaffection seems to have existed in the Philippine Military Academy for

many years. See *Far Eastern Economic Review,* 28 December 1989, p. 19.

37. Ambassador Platt to Representatives Renato Unico and Ramon Bagatsing, Jr., in Manila, as cited by *National News,* Manila, 12 October 1988, pp. 1, 10.

38. Singapore's offer was something of a political bombshell within ASEAN. Indonesia and Malaysia were quick to officially frown on it. However, extensive U.S. Navy use of Singapore ship facilities, including repair and resupply, has been a fact for many years. Singapore and Washington have been conducting feasibility studies on an expansion of American access. See *Far Eastern Economic Rview,* 17 August 1989, p. 12, and 31 August 1989, pp. 9–10.

39. Gorbachev at Krasnoyarsk, 16 September 1988, as cited in a Tass *Pravda* report. See FBIS-SOV, 19 September 1988, pp. 43–44.

40. Seth Mydans, "Moscow Hints at Leaving Cam Ranh Bay," *New York Times,* 23 December 1988, p. 3.

41. Always persistent, despite their public denials, the Soviets signed a trade protocol with Manila officials in March 1989 that grants Moscow permission to repair up to 120 of its fishing vessels at Cebu and Batangas shipyards. This was the first agreement of its kind. Soviet Aeroflot landing rights are also being negotiated. *Manila Chronicle,* 19 March 1989, p. 1.

42. See Lawrence E. Grinter and Young Whan Kihl, eds., *East Asian Conflict Zones: Prospects for De-escalation and Stability* (New York: St. Martin's Press, 1987), concluding chapter.

EIGHT

Security in the Southwest Pacific

Owen Harries

Introduction

Over the last half century the prominence of the Southwest Pacific[1] in world politics has fluctuated remarkably. For a brief interlude in the early 1940s, the islands and seas of the region were transformed with great suddenness from obscurity and insignificance into vital strategic objectives. Almost overnight, locations like Guadalcanal, Tarawa, Buna, and the Coral Sea became the focus of intense attention as two great powers fought fierce battles to control them, with vast stakes turning on the outcome.

The moment passed, and for the next several decades the region lapsed back into geopolitical obscurity. While most other parts of the world were affected in one way or another by the cold war and the upheavals accompanying and following the dissolution of empires, this area remained substantially untouched and inconsequential. It was stable, peaceful, peripheral and uncontested. Global strategists, with much else to occupy their minds, paid it scant attention, and on the few occasions it rated a mention, it was usually (and rightly) dismissed as a strategic backwater.

Yet even as this description was being applied, the status of the region began changing once again—much less dramatically and radically than in the 1940s, to be sure, but still quite perceptibly. Since the late 1970s, and especially since the mid-1980s, not only the two superpowers but a number of other major countries (France, Japan, China) have begun to pay it more attention and to formulate new policies in relation to it. Distant countries with no previous history of involvement—notably Libya and Cuba—have been dabbling in its affairs. Events within the region itself—discord among allies, anticolonial violence, bilateral agreements with hitherto excluded outside powers, intraregional declarations with strategic implications, and military coups in one leading island state—have attracted world attention, though the renewed interest also reflects changes beyond the region. One must not exaggerate; the visibility the Southwest Pacific has achieved during this new phase has so far been modest, and may well remain so. But a change there has been, and, for better or worse, the Southwest Pacific is becoming "interesting."

Strategic Significance of the Southwest Pacific

It is worth reflecting on this sequence of changes in the region's political and strategic status, for it bears on the general question of what can confer prominence on parts of the world that, on the face of it, appear to be of little consequence. Not only skeptical outsiders, but many inside the Southwest Pacific region itself find it difficult to answer the questions: How could this part of the world acquire serious strategic significance? Is it not too weak, too poor, too sparsely inhabited, too lacking in strategic assets, too marginal to become an arena of consequence in the affairs of nations?

One way of answering these questions, and the most readily resorted to, is to list the actual and potential strategic "assets" of the region: the fact that important sealanes of communication pass through it; that it offers a vital alternative route between the Pacific and Indian Oceans should the straits of Southeast Asia ever be interdicted; that its waters are rich in fish and that the mineral wealth of its sea-bed may in the not-very-distant future be of enormous value; that some of the islands could constitute very useful listening-posts for intelligence purposes.

While one could compile a reasonably impressive list along these lines, it has to be conceded that it would be less than compelling. Such a list would be likely to do no more than convince the doubting that the region is not entirely inconsequential. More to the point, such a concentration of attention on a region's actual or potential strategic assets encourages a fundamental misunderstanding of the nature of strategic importance. It cannot explain why the salience of a region may vary enormously even over a short period, and gives no answer to the question of why places that are strikingly deficient in such assets—a Fashoda in the 1890s, a Tarawa in the 1940s, a Grenada in the 1980s—can assume great importance with disturbing regularity.

There is another way of coming at the matter, one given point and credibility by the recent historical experience of the Southwest Pacific. Strategic importance is not something that inheres in a country or a region as an immutable, fixed quality. It is seldom derived principally from the characteristics of the region itself, and sometimes has little to do with them. It is contingent and situational in character, rather than intrinsic; even features that are superficially incontestably and inherently important in their own right (a "choke-point," say, or reserves of a "strategic" mineral) are only significant given certain assumptions about the intentions and motives of major actors on the international scene (and, in the longer run, about the persistence of certain technologies).

Fundamentally, it is the patterns created by the competing wills, intentions, and capacities of these major actors that confer or deny strategic importance, and the characteristics of regions signify as "assets" only insofar as they interact with and are caught up in those patterns. It is to those interactions, therefore, not to regions as entities-in-themselves, that we should primarily direct our attention in trying to anticipate what will appreciate and what decline in strategic relevance. The relevant lesson of World War II is that the importance of the Southwest Pacific between 1941 and 1945 had little to do with the internal characteristics of the region; the key was to be found in Tokyo and Washington, not in the Solomon Islands or New Guinea. Likewise, the key to understanding the importance of, say, Angola in the 1980s is to be found in the interaction of superpower rivalry and the issue of apartheid, not in the terrain and resources of Angola. It would be wise to keep this in mind in thinking about the Southwest Pacific today.

One further point. In speculating about the way in which a region like the Southwest Pacific is likely to figure in the affairs of great powers, it is as well to remember the stress that both practitioners and theorists of strategy have placed on indirection and nonlinear thinking.[2] The shortest route from A to B (in either the geographical or the political sense) is usually not the most strategically efficacious one; the long, less obvious, and apparently more difficult way around is often better. If a step is obviously logical, avoid it and look for alternatives, for it will have been anticipated and guarded against. The validity of the "indirect approach" to strategy helps explain the importance that peripheral and seemingly unimportant parts of the world regularly assume in the affairs of great powers. Paradoxically, the very fact that they are widely considered to be comparatively unimportant may enhance their attractiveness as objects of strategic attention, in the search for an unexpected and therefore effective point of entry and purchase.

Closely related to the concept of indirection is that of displacement. Powers sometimes conclude that conducting their rivalry in the area where their interests actually compete most sharply is too dangerous. The situation there is so sensitive and tense as to make the risk of armed conflict unacceptably high in the event of any strong move. In such circumstances the active competition is sometimes removed from the area of primary importance to a region where the stakes are lower and where there is greater freedom of maneuver. Thus at the end of the nineteenth century, European great power rivalries were played out largely in Africa and the Middle East (a displacement deliberately encouraged by Bismarck). More recently, the cold war antagonists looked for third world venues to conduct a struggle that was essentially Euro-centered. Bearing in mind the emerging multipolar system in the Pacific, this phenomenon of displacement is worth remembering in considering the relationship between the Northwest Pacific, where the stakes are very high, and the more peripheral Southwest Pacific.

Decolonization and Its Effects

The insignificance of the Southwest Pacific in world politics through the 1950s and 1960s was a function of two sets of conditions: the region was firmly under the control of the United States and its allies (Great Britain, France, Australia, and New Zealand); and America's global rival, the Soviet Union, lacked the effective reach and capacity to do anything about that control. As a result, the competition and conflict that give political and strategic salience to a region were absent. Strategically, it was an "American lake."

In the 1970s, the conditions that sustained this state of affairs began to disappear. First, the major defeat suffered by the United States in Vietnam and the simultaneous impact of Watergate destroyed the aura of invincibility and permanence that had hitherto surrounded America's dominance in the region (in much the same way, though not to the same extent, that the fall of Singapore had destroyed the myth of British supremacy a generation earlier) and created a domestic climate in America that inhibited the use of United States force in any circumstances short of an overt attack on clearly vital interests. Thus both the perception and the reality of American power were affected. While American influence in the Pacific continued to be great, it no longer had the aura of a force of nature that it was useless to challenge.

Second, in the 1970s the decolonizing winds of change that had already swept through Africa and Asia finally reached the Southwest Pacific in full force. The British, having already decided to withdraw from east of Suez in the late 1960s, had no reason to hold onto their scattered possessions in the Pacific. An Australia that was sensitive to third world pressure and to the example of London hastened to follow suit. Only France held on, but it did so against the tide. Between 1970 and 1980, seven new states came into existence in the region: Fiji, Tonga, Papua New Guinea, Tuvalu, the Solomon Islands, Kiribati, and Vanuatu.[3]

Except in the case of the last of these, where there was some externally inspired violence at the time of independence,[4] the decolonizing process was generally amicable and relations between erstwhile masters and subjects remained good. Nevertheless, the change was real. The new states had the normal sensitivities and concern over status of the newly arrived. They took their independence seriously and were concerned to establish that it meant something. For the first time the islanders had the right to deal directly with third powers, without consulting or receiving permission from anyone—and, conversely, they were more accessible to other states than they had been. Given that some of the Western colonizing states had grown accustomed to taking their domination for granted and to imposing their will on the region without bothering much about the wishes of local populations, there was a potential for friction unless some bad habits were quickly unlearned. That potential was increased by the fact that the move to independence coincided roughly with the emergence of a new generation of island leaders, ones with no personal memory of World War II and the close cooperation that had existed between the island people and the West during that period, but with some exposure to prevailing anti-imperialist (that is, anti-Western) and neutralist third world ideologies. As an astute and well-informed American diplomat observed, because of this generational change "the comfortable past, 'taken for granted' biases of most Pacific island states towards the interests of the ANZUS partners is becoming history."[5]

A third new factor that became increasingly significant as the 1970s drew to a close was the upgrading of the Soviet Union's interest and presence in the Pacific. This can be explained in terms of several interacting and mutually supporting factors: the humiliating experience of the 1962 Cuban Missile Crisis and the consequent Soviet determination to extend the global reach of Soviet power, in order to avoid any repetition; the comparatively good performance of the Soviet economy at the time, which made a sustained military buildup of unprecedented scope feasible; the opportunities offered first by the British withdrawal from East of Suez in 1968 and then by the semiparalysis of United States foreign policy in the mid-1970s; and, certainly not least, the increasing concern in Moscow about the long-term threat posed by China. (By the late 1960s and early 1970s this was acute. According to William Hyland, who occupied key positions in the Nixon and Ford administrations and dealt with the Soviets at the highest levels, China was "Brezhnev's obsession" during these years.[6])

The most obvious evidence of this new interest was the priority given to building up the Soviet Pacific fleet and the development of a patron-client relationship with Vietnam. In the twenty years between 1965 and 1985 the Soviet Pacific fleet doubled in size and, with over 800 ships, became the largest of the four Soviet fleets. Improvement in quality of equipment kept pace with growth.[7] As for Vietnam, the Soviet Union concluded a treaty of friendship with that country in 1979 and showed a willingness to subsidize its failing economy and to underwrite its invasions of its neighbors. In return, a Soviet military presence was established in Cam Ranh Bay, which by the mid-1980s had become the largest Soviet base outside the Warsaw Pact countries.

The main Soviet interest was, of course, focused on the Northwest Pacific and Southeast Asia, but it was not entirely restricted to those regions and there were good reasons for paying increased attention to the Southwest. While for some purposes it may be convenient to divide the vast Pacific region into subregions, it is a single entity containing no natural barriers or lines of demarcation.[8] To the extent that parts of the region assume greater importance, contiguous and flanking zones also assume a new interest, especially for those thoroughly familiar with the logic of the "indirect approach" and trained to think in terms of connections, themes, and long-term trends rather than in issue-specific and reactive terms.

Moreover, the Southwest linked with another theme of Soviet foreign policy in the 1970s. In addition to its developing interest in the Pacific, Moscow adopted a much more activist and ambitious approach to the third world at this time. While it centered mainly on Africa and the Middle East, this strand of Soviet policy would also naturally have aroused interest in the possibilities suggested by the emergence in the Pacific of a cluster of new, small, inexperienced, and poor states (but ones which, thanks to the Law of the Sea, claimed sovereignty over huge stretches of ocean). The fact that China was active in developing relations with these states acted as a further stimulant.

Another factor that must have attracted Soviet attention to the southern part of the ocean in the early 1970s was the coming to power of Labor parties in both Australia and New Zealand. In the case of Australia, the Whitlam government (1972–75) ended a period of over twenty years during which Australia had been governed by a conservative coalition that was strongly anticommunist and unwavering in its allegiance to the American alliance. Although the new government remained loyal to ANZUS, it generally deplored an assertive anticommunism as "obsessive" and "outdated" and laid stress on the need for an independent foreign policy reflecting Australia's "national identity." Some of its senior members were clearly and openly hostile to the United States. These developments were of obvious interest to Moscow. Indeed, taken in conjunction with the effective demise of SEATO (the South East Asian Treaty Organization) and the British entry into the Common Market, Dalton West perceptively observes that, from a Soviet perspective not only the United States but Australia and New Zealand themselves were in a kind of psychological "decolonizing mode" at this time and therefore worthy of Soviet attention.[9]

Along with the three developments mentioned above—the

effects of defeat in Vietnam on United States policy and prestige, the decolonization of the region, and the upgrading of Soviet interest—one other factor worked to enhance the region's standing in the 1970s. By the end of the decade the sustained and spectacular economic growth taking place in the Northwest Pacific and parts of Southeast Asia was attracting attention to the region generally, with much talk of the "shifting of the center of gravity of world politics" and the "coming Pacific Century." Japan was clearly on the verge of becoming an economic superpower, and who could tell what political and strategic implications this would have in the not very distant future? In Japan's wake, the "Four Tigers"— South Korea, Taiwan, Hong Kong, and Singapore—were making spectacular progress that no one had foreseen a generation earlier; again, who could predict to what other improbable places in the region this dynamism might spread? Concepts such as "the Pacific Community" and "the Pacific Rim," which were gaining a good deal of currency, tended to be inclusive in nature, encompassing the vast stretches of the South Pacific as well as the industrial dynamos of the north. To some extent, therefore, a rising tide lifted all boats: the Southwest might be one of the least important subregions of the Pacific; but if the Pacific was destined to become the most important region in the world, even its remoter parts acquired a new significance.

Internal Security Threats

If the developments described above set the contemporary stage in the Southwest Pacific, how have things played out in the 1980s?

The decline in the authority of the United States following the Vietnam defeat did not last. During the Reagan administration there was a recovery of confidence and assertiveness, manifested in the Pacific by the buildup of naval power and the adoption of an assertive maritime strategy. But while they lasted the demoralization and loss of prestige had important consequences. They contributed to a perceptible strengthening of anti-American sentiments in the region (in Australia and New Zealand, as well as in the islands) and were probably a necessary, though not sufficient, condition for the crisis in the ANZUS alliance—America's linchpin regional security arrangement—that was precipitated by New

Zealand's refusal to permit continued visits of United States nuclear-powered or nuclear-armed vessels to its ports. That crisis, in turn, contributed significantly to an atmosphere of uncertainty and instability in the affairs of the region as a whole and encouraged other challenges to American policies and authority along quasineutralist, antinuclear lines.[10]

Given the responsiveness of American governments to internal pressure groups and the historical insensitivity of Washington to the interests of the Pacific islands (Micronesia as well as the Southwest), it is likely that in any circumstances the United States would have been slow to respond to complaints about American fishing of tuna in the Exclusive Economic Zones of the islands. But, again, the American post-Vietnam trauma, which manifested itself particularly in a neglect of the Pacific, did not help. In any case, the fishing issue was allowed to fester for several years and contributed substantially to a souring of relations between the United States and the newly independent island states before it was finally resolved in 1987.

As for the new states themselves, for a number of years following independence things went reasonably well. They enjoyed internal stability and order under civilian governments, and there was a gratifying absence of coups, military interventions, and the other manifestations of social discontent and political cleavages experienced by many third world countries in the postindependence period. As for relations among the new states, here again there was little evidence of discord and much stress on an alleged tradition of cooperative and harmonious modes of behavior, characterized as "the Pacific way." As early as 1971, the Pacific Forum had been formed as a regional organization (it included Australia and New Zealand as well as the island states) to facilitate consultation and regional cooperation, and in a low-key and modest way it seemed to do the job. At the beginning of the 1980s there was a general air of self-congratulation at the way things were going, and flattering comparisons were often drawn between the region and other more troubled parts of the third world.

But as the decade advanced it became clear that all was not as well as it seemed. Most of the countries of the region were poor and, given their paucity of resources and high population growth rates, destined to become poorer. Their independence and freedom of action were compromised by a heavy reliance on aid, mainly from Australia and New Zealand (per capita, the region receives more aid than any other in the world). It did not take long for it to become apparent that, both in the internal affairs of individual states and regionally, the capacity for harmony claimed for "the Pacific way" was grossly exaggerated. By the late 1980s there was increasing evidence of discord and conflict. Papua New Guinea, the largest state of the region, was characterized by systemic corruption, political incoherence, instability, crimes of violence, and a secession-threatening regionalism. In 1987, Fiji, the second most important state, suffered two military coups reflecting acute communal tensions between the Fijian and Indian components of the population. In 1988 there was destructive rioting in the capitals of Vanuatu and French Polynesia, signs of increasing political and economic frustrations. Regionally, the apparent unity represented by the Pacific Forum was belied by evidence of differences between the Melanesian islands of the west and the Polynesians of the east; in 1986 Papua New Guinea, Vanuatu, and the Solomon Islands formed a "Melanesian Spearhead" group, intended to promote the independence of New Caledonia but also to give the Melanesian states more clout in what they regarded as the "clubbish" atmosphere of the Forum. This in turn has led to the creation of a countervailing "Polynesian economic summit."[11]

As well as from the problems of postcolonialism, the security of the region has also suffered from two unresolved cases of colonialism: in French New Caledonia and Indonesian West Irian.[12] New Caledonia represents a genuinely difficult problem (in that the indigenous Kanaks constitute only 43 percent of the population) but has been made much worse by French mishandling, at least until very recently. While there is no force in the region that is capable of mounting a serious challenge to the French (who have more than 6,000 security personnel in the territory), the issue has served to fuel anti-Western feeling, to create precedents for bloodshed, and to present an opportunity for outside meddling in the region's affairs (particularly by Libya). Recent steps taken by the Rocard government to establish an acceptable route for constitutional evolution may help defuse the issue, but much damage has already been done.

The West Irian issue is important because of the tension it has caused between Indonesia and Papua New Guinea since the former Dutch territory was incorporated into Indonesia in 1969.

The Melanesian opposition to Indonesian rule, the Organisasi Papua Merdeka (OPM), makes free use of Papua New Guinea as a sanctuary, and border incidents involving Indonesian troop incursions across the frontier have occurred frequently (on at least three separate occasions in 1988). There is considerable sympathy for the Melanesian cause in Papua New Guinea and considerable Indonesian resentment of that sympathy and the resulting tolerance afforded the OPM.

While the OPM is neither a coherent nor a formidable force, it could conceivably one day produce a serious crisis between Indonesia and Papua New Guinea, particularly if President Suharto's eventual departure should result in more volatile leadership in Jakarta. This in turn could involve Australia, which has close ties with Papua New Guinea and which is committed to consult "for the purpose of each Government deciding what measures should be taken" in the case of external attack.[13]

The West Irian issue apart, it is difficult to see how the various sources of tension discussed above could, *in themselves,* create serious regional security risks; the distances involved are so great and the capacity to project power effectively is so limited as virtually to guarantee the localization of any conflict.[14] (West Irian is an exception precisely because it has the potential of involving regional actors—Australia and Indonesia—who *do* possess a significant capacity to project power.) The caveat "in themselves," however, is important. For while their potential for initiating purely regional conflict is very limited, some of these sources of weakness and tension could encourage and facilitate outside intervention and manipulation with serious security consequences.

In this respect, one feature of island life deserves particular attention. Consider the following cluster of facts: First, the island economies are poor and the scope for economic development is very limited.[15] Second, although populations are small, the rate of population growth is high (about 2.3 percent, which means a doubling of population in about thirty years, and a large component of young people). Third, the drift from the country to the towns is marked, particularly among the young; it is straining very modest urban infrastructures, creating urban underemployment, and in some instances (for example, Port Moresby) leading to a breakdown of law and order. Fourth, by third world—and indeed worldwide—standards the region has a very literate population

(Papua New Guinea apart, about 90 percent). Education—and the non-manual work to which it normally gives access, but not much of which exists in the islands—is highly valued, while the tradition-al occupations of cultivating the land and fishing are losing status.[16]

This is a dangerous set of conditions. Historically—and this applies both to Western and third world countries—the emer-gence of sizable groups of young, recently urbanized, underem-ployed, and educated (or semi-educated) people whose aspirations stand little chance of being satisfied has spelled politi-cal trouble. Such groups have proved easily mobilizable and manipulable; characteristically they have provided the frustrated and idealistic cannonfodder for extremist movements. Whether the youth of the island states will follow this pattern, or whether the extremely small populations and the highly personalized nature of island politics will make a decisive difference (by not providing the necessary "critical mass" and minimizing "alienation," respective-ly), remains to be seen. The answer is likely to provide a crucial indicator of the future stability and security of the region.

Superpower Rivalries

In the longer run, the fact that most of the newly emerging powers—Japan, China, India—are located in or near the Pacific is likely to be of great significance for the Southwest Pacific, if only in terms of the displacement effect discussed earlier. But in the immediate future the most serious danger to the security of the region is the possibility that it will become an arena for active superpower competition. For this to happen, the Soviet Union will have to make considerable progress in advancing its interests in the region.

The nature of those interests is often trivialized, with such things as fishing and the cruise ship business, both of little importance in themselves, given prominence. More is at stake than that. In the broadest sense, as long as it sees itself engaged in a global contest for primacy with the United States, the Soviet Union has an inter-est in making incremental gains in that competition wherever and whenever it can. Given that contest, the availability of capacity and opportunity—for a superpower either to advance its own position or to weaken that of its adversary—is in itself enough to create an interest. In recent years, an enhanced Soviet presence in the Pacific (as well as such relevant assets as huge merchant and fishing fleets) have provided an increased capacity; the period of weakened American authority and resolve, together with rapid decolonization, appeared to provide an opportunity. Moscow has not failed to attempt to exploit such circumstances elsewhere in the world when they have arisen, even in places where there was no record of previous Soviet involvement and no apparent specific interest. No special explanation, therefore, is required as to why the Soviets should choose to do the same in the case of the Southwest Pacific; such a choice is implicit in the global competition. In the early stages, the negative goal of undermining the American presence and weakening the links between the indigenous states and the West is likely to take precedence over the positive extension of a Soviet presence.

As already indicated, a greater interest in the Southwest Pacific is also implicit in Moscow's clear and repeatedly stated aim of establishing itself as a major and legitimate Pacific presence. (The most authoritative expression of that aim was, of course, Mikhail Gorbachev's Vladivostok speech in July 1986.) For, as was argued earlier, that aim will be advanced to some extent by increased acceptance and influence *anywhere* in the region; and if the more important countries of the Northwest Pacific are for the time being in the "too hard" category, then that is all the more reason for trying to make progress in the more marginal, but also possibly more amenable, countries further south. Moreover, while the states of the Southwest Pacific are small, they are also numerous; and Moscow understands well that, on some issues and in some contexts, numbers count in contemporary international politics. This is true of a whole range of existing international organizations, and should a "Pacific Community" or any other comprehensive Pacific regional organization be created, it is likely to be true of it also.

As for Soviet military-strategic interests, the most often mentioned include the potential of the deep ocean troughs and trenches of the region for deploying and hiding submarines, and for antisubmarine warfare (ASW). Soviet research on the hydrography and oceanography of the South Pacific is reported to be already more extensive than that of any other nation;[17] presumably closer ties with island states, leading to easier access and a more sustained presence, would facilitate that research. While in terms of the overall Pacific naval scene, in which the main forces and potential for engagement lie well to the north of the region, the waters of the Southwest are presently of limited importance, they would assume much greater significance in the case of the closing of the straits of Southeast Asia. Prudence, therefore, indicates the need for such attention on the part of the Soviets.

The facilitating of intelligence gathering—including easier surveillance of United States missile and SDI research in Kwajalein, in the Marshall Islands to the north, and the setting up of an array of underwater acoustic devices off the coast of Vanuatu—is also advanced as a Soviet interest, though some specialists express skepticism.[18]

According to recent expert opinion, Moscow (and for that matter, Washington) may have another and much more compelling strategic interest in the South Pacific. Aadu Karemaa, manager of advanced antisatellite systems at General Dynamics, writing in the *Proceedings* of the U.S. Naval Institute, observes that the Soviets' three "space ports," from which all their space launches—including those of their anti-satellite weapons—take place, are located directly on the opposite side of the earth to a triangle in the South Pacific. Any satellite launched in any direction from the Soviet space-ports will pass over this triangle after completing the first half of its orbit. This means, says Karemaa, that the South Pacific triangle constitutes the strategic space equivalent of a "narrow sea" in naval warfare. As Karemaa puts it, "Control of areas opposite space launch facilities could deny entry and exit to the respective space programs just as control of Gibraltar or the Straits of Hormuz could deny entry and exit to some critical ports."[19] If this is the case, the Soviet Union has a very serious interest indeed in the South Pacific, in order to ensure its future access to space by being in a position to contest United States control of that vital triangle. While the area in question is to the southeast of the island states, they offer the nearest possible support bases for such operations; and the American radar network to track newly launched Soviet satellites (which would be the basis of a United States anti-satellite system should one ever be acquired) spreads over the whole Pacific region.

Recent Soviet Activities

To promote the interests described above, the Soviet Union has followed a familiar two-track policy, combining attempts to strengthen its government-to-government relations in the region with efforts (conducted through both pro-Soviet agents and inde-' pendent actors who hold views congenial to Moscow) to create and manipulate local organizations that can advance its cause.

At the governmental level, Soviet efforts have included initiatives to establish a stronger diplomatic presence, offers of collaboration in oceanographic research, fishing agreements, proposals to develop trade and other forms of economic cooperation, and the arranging of visits to the Soviet Union and other Eastern bloc countries. As well, the Soviet Union has made a point of expressing regular support and sympathy for a range of policies adopted by the governments of the region, particularly those of an antinuclear, neutralist, and anti-"imperialist" character. It has enthusiastically applauded the ban on visits by United States nuclear-armed or -powered ships imposed by New Zealand, Papua New Guinea, and the Solomon Islands; welcomed the declaration of a South Pacific Nuclear Free Zone (SPNFZ) and ratified the protocols of the treaty (which the United States, Britain, and France have not done); and supported the cause of New Caledonia's independence at the UN and elsewhere.[20]

The progress made along this conventional diplomatic track has so far been fairly modest. Although a number of countries (Western Samoa, Tonga, Fiji) have granted nonresidential diplomatic accreditation to the Soviet Union, only Papua New Guinea has agreed to the establishment of a permanent mission. The Soviets entered into fishing agreements with Kiribati (1985) and Vanuatu (1986), but both agreements subsequently lapsed when the Soviets attempted to alter their terms. The invasion of Afghanistan in 1979 thwarted an oceanographic research proposal (and ended Soviet cruise ship visits to the islands for a number of years).

One of the reasons for this limited success is that Soviet moves at the governmental level are normally highly visible and have tended to evoke responses from Western governments. Thus an early Soviet approach to Tonga for a fishing agreement in the mid-1970s was quickly followed by a quadrupling of Australian aid to the region; more recently the Soviet fishing agreements with Kiribati and Vanuatu served to concentrate Washington's attention wonderfully and led to a speedy resolution of the tuna fishing dispute. Moreover, though the island leaders are for the most part inexperienced in international affairs, they have not been particularly naive in assessing Soviet initiatives. While prepared to pocket support from Moscow for causes they have adopted of their own volition, they have been careful about taking any steps that would promote superpower rivalry in the region.

What success the Soviets ultimately have diplomatically will depend to a considerable degree on how much progress they make along the second track in developing influential constituencies for themselves in the islands, in the form of either outright front organizations or among groups that are disposed to friendship on principle. The two main targets-cum-instruments of penetration in this respect are labor unions and churches (with institutions of higher education close behind). The first of these—the unions—represent a classic communist target. Throughout their history and all over the world, communist parties have vied with social democrats for control over organized labor, the natural political base for movements of the left in modern societies. The churches are a newer target, reflecting recent changes in the social doctrines of Christian churches as well as a shift in the balance of power in the largest existing ecumenical organization, the World Council of Churches.

The decision to pursue a more active and aggressive policy towards the fledgling unions of the Southwest Pacific may be traced back to the conference of the communist-dominated World Federation of Trade Unions (WFTU) in Prague in 1978. The chosen instruments were leading figures of the hard left, pro-Soviet elements in the Australian and New Zealand union movements, many of whom were present at the Prague conference. They subsequently played a prominent role in the creation of the transregional Pacific Trade Union Forum (PTUF) in 1980, the key union organization in promoting pro-Soviet and anti-Western causes.

The PTUF/C[21] has been notable for the degree to which it has subordinated purely labor, work-related issues to broader political ones, and for its openly selective treatment of those issues. It has opposed Western military bases—the United States bases in the

Philippines and intelligence facilities in Australia—but has been silent about the Soviet base in Cam Ranh Bay. It has attacked Western "imperialism" but had nothing to say about the Soviet invasion of Afghanistan. Likewise, its support for the aboriginal minorities of the world stops short when it comes to the treatment of the Miskito Indians of Nicaragua by the Sandinista regime. Consistency is to be found not in the treatment of particular issues by the PTUF/C, but in its ready support for those causes favored by the Soviet Union: opposition to the ANZUS alliance, the strengthening of antinuclearism in the region, anti-Westernism in the guise of opposition to imperialism, and neutralism—either in the form of balancing the presence of the two superpowers in the region (by enhancing that of the Soviet Union) or by excluding them both (which would diminish the American presence).

As well as pushing these causes, the hard left elements in the PTUF/C have made strenuous efforts to discredit attempts by moderate American and Australian unionists to create countervailing regional organizations, usually by denouncing them as CIA inspired. Thanks largely to the help given by some Australian and New Zealand journalists with access to leading newspapers, radio, and television, they have achieved considerable success in this respect.[22]

As for the use of the Christian churches to further Soviet interests in the region, this must have come as a surprise to many observers. For until quite recently it was quite usual to maintain that because Christianity was strong in the islands (as a result of intensive missionary activity in the colonial period) the region was insulated against radicalization, particularly of the communist, atheistic variety. Such confidence was doubly misplaced. In the first place, historically the Christian churches have played a radical, and even revolutionary, role as often as they have played a conservative one. (In sixteenth- and seventeenth-century Europe, for example, churches provided a major vehicle for assaults on the established order by what would now be called counter-elites representing a counter-culture.) Second, in the 1970s and 1980s that confidence was particularly inappropriate as it overlooked the profound changes that have occurred in the Christian churches with the rise of "liberation theology" and the social and political doctrines associated with it. During the same period, the most powerful ecumenical organization, the World Council of

Churches, has also changed its character radically as the representatives of Eastern bloc and third world churches have come to dominate its affairs.[23]

The WCC encouraged the formation of the Pacific Conference of Churches in the 1960s and the Independent and Nuclear Free Pacific movement in the 1970s. Together with the PTUF/C, these organizations have played leading roles in developing and spreading anti-Western themes in the region. The WCC has provided financial support, helped organize conferences, brought in liberation theologians from Europe, Latin America, and Africa, and worked directly to support favored political parties (such as the Vanuaaku Party of Vanuatu).[24]

By the mid-1980s the Australian foreign minister, Bill Hayden, was moved to describe "a surprisingly high level of Soviet-backed activities in South Pacific countries."[25] It involved at least loosely coordinated action by unionists, churchmen, and educators working through a network of organizations. How effective all this effort has been, and how far it has advanced Moscow's interest, is difficult to judge at a distance. In a well-informed and judicious assessment, Michael Easson, a moderate Australian union official with a background of work in the region, has argued that the degree of success achieved by the left with respect to labor unions has been exaggerated frequently, partly out of a concern to draw attention to the problem.[26] Easson maintains that it would be "simplistic and unduly pessimistic to argue that the PTUF/C has reached an unassailable position of influence in the region or has converted the Pacific labor movement to an anti-Western policy posture," and that PTUF/C conferences "provide a misleading view of the total scene." In support of this view he points to the assistance provided to Pacific unions from noncommunist sources, including Denmark, the Israeli Histadrut, the United States AFL-CIO, and the Commonwealth. He also cautions against attaching too much importance to the attendance of senior Pacific island labor leaders at WFTU conferences in Eastern Europe, arguing that in many cases this is an indication of a thirst for overseas travel rather than of ideological conversion.

While the points are well taken, there is little room for complacency. Communist tenacity and patience in pursuit of a desired goal should not be underestimated, and neither should the advantage deriving from clarity of objectives. (While much of the

Western assistance to the region's unions goes to supply such mundane things as desks and typewriters, the left concentrates singlemindedly on political goals). One instance of persistence that paid off is Papua New Guinea's yielding to pressure and agreeing in principle in 1988 to a permanent Soviet diplomatic mission in Port Moresby; now that the precedent is established, others may follow. Further, the pro-Soviet elements have the advantage that they do not have to work to manufacture sentiments that are alien to the Southwest Pacific, but only to strengthen, organize and manipulate ones that are already spontaneously held. In a region that over the last forty years has been used quite carelessly and often disgracefully by Western countries to test nuclear weapons, antinuclear attitudes existed before the left began to exploit them.[27] The same is true of anticolonialism, a concern with native rights, land rights movements and environmental issues. These views are not restricted to fellow travelers or those described by Lenin as "useful idiots."

Another development that has created conditions favorable to the Soviets throughout the region is the virtual destruction of ANZUS as a functioning alliance following the New Zealand decision to ban visits by nuclear-powered and -armed ships to its ports. That decision was taken against a background of strengthening "peace" and environmental movements in New Zealand, and, despite claims to the contrary, a growing anti-Americanism (which often takes the form of an assertion of "moral equivalence" between the United States and the Soviet Union). As recently as March 1989, a staff report on the South Pacific presented to the Committee on Foreign Affairs of the U.S. House of Representatives expressed concern that "New Zealand might ultimately move in the direction of neutralism or even isolationism."[28]

As New Zealand is a very significant presence in the Southwest Pacific, its position has had repercussions throughout the region. Instead of dealing with three Western countries having shared views and policies, the political elites of the island states now find that one of the countries to which they are accustomed to look for guidance is adopting and defending policies that are praised by the Soviet Union. From the point of view of the islanders this means that they now have a choice between two conflicting Western policies. From Moscow's point of view it means that their efforts to propagate neutralist and antinuclear policies are legitimized by one of the two significant Western countries of the region.[29]

Last, in assessing the prospects of Soviet success in the Southwest Pacific one should bear in mind what was said earlier about the emergence of a frustrated and alienated stratum of young people, underemployed and at least partly educated, in the island states. This, together with the paucity of independent associations in these minute societies, gives the existence of a few well-organized and purposeful radical bodies much more significance than it would have in larger, more complex, and more prosperous societies. In the end, and despite his proper concern to keep the pro-Soviet activities in perspective, Easson concedes that the PTUF/C has become "the only major forum for regional union conferences and regional meetings of Pacific union leaders" and that it sets the agenda on many issues in the region.

To round off this section on pro-Soviet activities in the region, something needs to be said about the activities of outside countries that have good relations with Moscow. The most publicized case is that of Libya. Muammar Gadhafi's regime is known to have given military-political training to two groups of Kanaks from the small *Front Uni de Liberation Kanak* (FULK) party in New Caledonia and to a small number of ni-Vanuatu, the indigenous people of Vanuatu. It has also established links with the OPM of West Irian and attempted to establish a Libyan People's Bureau in Port Vila, the capital of Vanuatu. While these activities have considerable mischief-making potential, it is doubtful that they were Moscow-inspired or that they have served Soviet interests. Gadhafi has probably been motivated mainly by a concern to pay back the French for their opposition to his North African policies, and there is little evidence to support the view that he takes instructions from Moscow. In any case, the principal results of his activities have been to raise the alarm level and to strengthen the case of those who are concerned about trends in the region. The same is true of Cuba's dabbling, the most noticeable expression of which has been the sponsoring of Vanuatu's membership in the non-aligned movement.

Of potentially more significance than either of these is the possibility—and it is no more than a possibility—that India may be interfering in the affairs of Fiji. In 1988 there were allegations by some American and Australian journalists that an illegal arms

shipment seized by Australian customs in Sydney while on its way
to Fiji originated in India. The sixteen-ton container—which held
submachine guns, Soviet AK-47 assault rifles, hand grenades, gre-
nade launchers, mortars, and antitank mines—was addressed to an
Indo-Fijian businessman and marked "Used Machinery." New
Delhi issued strong denials and the Australian government said
that it had no evidence of any Indian government complicity, but
United States and Australian journalists continued to maintain
that Australian intelligence agencies were following up "strong
circumstantial evidence" of the involvement of the Indian government.[30] Should these allegations turn out to have substance their
implications would be serious. Like Libya and Cuba, India has
close relations with the Soviet Union; but unlike them it has a
formidable navy and the capacity to project significant power into
the Pacific. Given its "non-aligned" interest and status and the
Indian majority in Fiji, it is not inconceivable that India could one
day decide to become a serious player in the Southwest Pacific,
perhaps in cooperation with the Soviet Union.

The preceding analysis has assumed continuity in Soviet policy
towards the Southwest Pacific, an assumption that many would
challenge at a time when it is widely believed in the West that
under Mikhail Gorbachev the whole foundation of Soviet foreign
policy is changing. With the retreat from Afghanistan and recent
developments in Eastern Europe, the Brezhnev Doctrine ("once
communist, always communist") has been abandoned. Even more
basically, some Soviet statements seem to indicate that the fundamental doctrine of international class war—the doctrine that
initially brought about the cold war—is being given up.
Gorbachev's regular and dramatic offers of arms reductions and
his talk of one European home fuel the sense of far-reaching
change.

But all that having been said, there are still reasons—some
general, some pertaining to the region itself—for not simply assuming that Moscow has lost all interest in the Southwest Pacific.
In the first place, it is worth bearing in mind that, whatever
Gorbachev and his colleagues are saying about a less confronta-
tional foreign policy and whatever trouble the Soviet economy is
in, the production of arms still remains at—or perhaps even
above—the enormously high level of the Brezhnev days. Second,
whatever Gorbachev's intentions and however genuine he may be,

the odds on his succeeding, or on his even remaining in office for very long, are not particularly good. Third, whether he endures or not, the Soviet Union will remain a global power with substantial interests in the Pacific, and, after all, it is Gorbachev himself who has made the definitive statements about the seriousness of those interests. As multipolarity becomes a reality in the region and as the Soviet Union's place in the international pecking order comes under increasing threat, it is surely more likely that the Soviet leaders—still with enormous military power at their disposal —will struggle hard to compete rather than bow out gracefully. Finally, it is worth noting that while Moscow has abandoned Eastern Europe, it has not yet shown a similar willingness to withdraw support for its various third world clients. As of the beginning of 1990, Soviet military aid is still pouring into Nicaragua, Afghanistan, Ethiopia, and Cambodia. This may, of course, change, but it cannot simply be assumed that it will, on the basis of what has happened in Eastern Europe.

As far as the Southwest Pacific in particular is concerned, the way things are shaping up may mean that, by a process of elimination, Moscow will find the region more rather than less attractive as a sphere of operation. The Soviet Union has committed much of its resources and effort to developing its position in the Pacific over the last twenty years, and presumably there are powerful groups in the Soviet system that would be strongly opposed to writing off that political investment without getting some returns on it. But at the same time, the priority currently given economic reform in Moscow means that the regime is both very cost-conscious in its approach to foreign policy and concerned to promote detente with the United States and other Western countries. The free-spending and tension-creating policies of the 1970s are, for the time being at least, out of fashion.

Given these conditions, the Southwest Pacific must appear a rather attractive option. First, it is in a more fluid state than other parts of the Pacific and offers greater prospects for easier gains that will save face and justify the investment. Second, it is a cheap area of operations where small inputs go a long way. While the Vietnams, Nicaraguas, and Cubas of the world eat up billions of Soviet dollars, the Southwest Pacific is a region where a few million can have a substantial impact (the fishing agreement with Vanuatu, for example, cost less than $2 million annually). Third, as it is not a high-risk area involving clearly perceived vital interests, incremental gains by Moscow are unlikely to cause a serious heightening of superpower tension that would conflict with the goal of detente. Precisely because it is not a region of vital strategic importance, and because the United States is likely to be more tolerant of Soviet activity there than it would be of moves in more sensitive regions, the Southwest Pacific is a suitable area to focus on during a period of relaxed tension.

If this is indeed a line of thought that some Soviet policymakers are likely to follow, the conclusion to be drawn by Western countries interested in the region is a familiar, if somewhat paradoxical, one. The less seriously the region is taken, the more likely it is to become a security problem; conversely, if it is treated as an area of some importance and attended to properly, it is likely to remain of modest significance. Despite the weakening of Western influence in the last decade, as of now the United States and its allies still enjoy great advantages. But if recent trends are allowed to continue over the next two decades, this would no longer necessarily hold true.

United States Policy Options

What is required of the United States in order to meet the security problems of the Southwest Pacific is not particularly difficult or expensive in money terms. The main dangers are American complacency and neglect; the main requirement, intelligent and sympathetic attention.

Any use of force by the Soviet Union in the near future is very unlikely and could be met more than adequately by the United States and its allies. For that reason, and because there is little or nothing that the island states can do about the military situation, it is a mistake for the United States to emphasize strategic and military factors in dealing with the islanders. Uniformed admirals armed with wall charts and CINCPAC lectures about Soviet submarines are not the appropriate emissaries to represent America: they serve only to arouse suspicions and fears—of the United States more than of the Soviets. The proper recipients of strategic messages are not the regional governments but the Soviet Union, which should be disabused of the idea that the United States will be tolerant of incursions into the region and that the pickings will be easy there.

As far as the islands themselves are concerned, the main problem is political and economic, not strategic. Both island governments and elites must be convinced that their future security will be best guaranteed by staying with their Western friends, not by neutralism or by engaging in the dangerous game of trying to play off one superpower against the other. There are some issues on which the United States cannot give way because larger interests are involved. These include attempts to denuclearize the region and to end the French presence there. If Washington has no choice but to remain firm on such questions, that is all the more reason for the United States' being accommodating and sensitive in other respects.

The appeals of neutralism, the antinuclear movement, and opposition to "imperialism" derive largely from a generalized and diffuse anti-Western (and particularly anti-American) sentiment. That, in turn, reflects a hypersensitive resentment and frustration at being taken for granted or slighted in other ways. To some extent such feelings are inherent in the island condition, a function of isolation and smallness. But they have been unnecessarily exacerbated by clumsy and unthinking American behavior. If the United States were to decide that the future of these islands is of little concern, such behavior—and the underlying attitude that these small countries are inconsequential and rather tiresome—would not matter. But as long as it thinks otherwise, the United States should act accordingly.

Attention and respect would go a long way towards dissipating the anti-Americanism that has grown in the region in recent years. The islanders need to be conversed with more and lectured to less. They need to be visited and invited. (If the Queen of England can manage to visit the region from time to time and to have the island leaders to dinner at Buckingham Palace, it should not be impossible for, say, an American vice-president to do the equivalent.) The United States should strengthen its diplomatic presence in the region, making full use of the small, dedicated group of diplomats who have labored over the years to acquire expertise in the affairs of this unfashionable region. Island leaders should be asked to attend regional meetings and encouraged to voice opinions. They should be praised for their achievements and helped with their problems.

Then there is the matter of United States aid. Currently, the United States provides a paltry $17 million a year, which amounts to about 2 percent of total assistance flows to the region. In symbolic terms as well as practically, this is insufficient. The region does not need huge quantities of aid, which would have a disruptive effect; what is given to Egypt or Israel every week—say between $35 and $50 million—would suffice for the whole Southwest Pacific for a year. Directed into well-designed projects through contractors (rather than into general budgets), such an amount could have an appreciable effect in alleviating the region's problems. It would also contribute to a much-needed improvement in America's image.

Former Secretary of State George Shultz was fond of using a gardening metaphor to describe his approach to foreign policy. It is particularly appropriate in the case of the Southwest Pacific, where careful cultivation is needed to make up for past neglect, and where the virtues called for are constancy of attention, patience, and respect for the nature of the soil and climate with which one has to work.

NOTES

1. In this article, the term "Southwest Pacific" will be used to cover Papua New Guinea (PNG) and the mini- and micro-states of the region. It will not include Australia and New Zealand.

2. See, for example, B.H. Liddell-Hart's very influential *Strategy: The Indirect Approach* (first published London: Faber and Faber, 1941), and Edward N. Luttwak, *Strategy: The Logic of War and Peace* (Cambridge, MA: Harvard University Press, 1987).

3. In 1965 the Cook Islands became self-governing, in free association with New Zealand, which retained responsibility for their defense and foreign policy. Niue followed in 1974. Western Samoa and Nauru achieved independence in the 1960s.

4. The French and the American Phoenix Foundation were the inspirers.

5. John C. Dorrance, "U.S. Security Interests in the Pacific Islands and Related Policy Issues," a paper delivered in Apia, Western Samoa, November 1988.

6. See William G. Hyland, *Mortal Rivals* (New York: Random House, 1987), particularly pp. 63–66.

7. See Alvin H. Bernstein, "The Soviets in Cam Ranh Bay," *The National Interest,* Spring 1986, for a well-informed account of the growth of Soviet naval power in the Pacific.

8. Cf. the observation of Australian Minister of Defence Kim Beazley, in his speech, "Australian Defence Policy," at the Bicentennial Conference on Australia and the World in December 1988: "[An] important consequence of recent strategic developments has been the breakdown of the traditional but in many respects artificial distinction which has been drawn between South East Asia and the South Pacific."

9. Comment to the author on this paper by Dalton West. If indeed the Soviets did interpret Australia and New Zealand in this way, subsequent developments indicate that they had considerable insight. In particular, New Zealand in the 1980s during the Lange Labor government (1984 to the present) has shown marked "decolonizing" characteristics, and the remark that it is "the first Third World country with good drinking water" is rather more than a smart crack.

10. The challenges also extended to other Western countries. In

1986, Prime Minister Walter Lini of Vanuatu described Australia and New Zealand as the biggest threats to the region's security, citing their military forces and their alleged neocolonialist ambitions. Lini later retreated from this position.

11. See David W. Hegarty's paper, "External Security Issues in the South Pacific," November 1988, pp. 8–9. The paper was delivered at a conference in Apia, Western Samoa. See also "Polynesia vs. Melanesia," *Islands Business,* Fiji, February 1988.

12. Both France and Indonesia would, of course, object to the characterization of "colonialism" and insist that in each case the territory involved is an integral part of the mother country. But in each case rule is imposed on indigenous populations that are different and distinct from those of the metropolitan populations.

13. *Joint Declaration of Principles Governing Papua New Guinea-Australia Relations* signed by the Prime Minister of Australia and Papua New Guinea in December 1987. See Hegarty, "External Security Issues," for a fuller account of the West Irian issue.

14. Only three of the island states—Fiji, Papua New Guinea, and Tonga—have regular defense forces. PNG, with Australian logistic support, did in fact lend support to the government of Vanuatu during the trouble in 1980. But the generalization still holds.

15. The clearest exceptions to the regional poverty are Nauru, which has large (but finite) phosphate deposits, and New Caledonia, the world's second-largest producer of nickel ore.

16. For the information contained in this paragraph see Robert C. Kiste and R.A. Herr, "The Potential for Soviet Penetration of the South Pacific Islands: An Assessment," December 1984. This paper was commissioned by the U.S. State Department. While Kiste and Herr provide this information, they do not relate these four aspects to each or attempt to draw a socio-political conclusion from them.

17. See Paul Dibb, "Soviet Strategy Towards Australia, New Zealand and the South-West Pacific," *Australian Outlook,* August 1985.

18. See David Hegarty, "The Soviet Union in the South Pacific in the 1990s," a paper given at a conference at the Australia

National University in May 1988, p. 11 of the draft version of the paper.

19. See Aadu Karemaa, "What Would Mahan Say about Space Power?," *Proceedings,* U.S. Naval Institute, April 1988. The author is the manager of Advanced Anti-Satellite and Anti-ASAT Systems at the Space Systems Division of General Dynamics Corporation.
   See also, Michael Richardson, "Why the Russians are Coming," *Pacific Defense Reporter,* September 1988, which follows up on Karemaa's article.

20. These activities should be related to simultaneous efforts by Moscow to develop stronger bilateral relations with Australia and New Zealand on a broad front, including trade, collaboration in space and medical research, and sport. Eduard Shevardnadze's visit to Australia in 1987 was the first ever by a Soviet foreign minister.

21. Its name was changed to the Pacific Trade Union Community in 1986, and the designation PTUF/C will be used here.

22. On the left and the Australian media, see the contributions of Michael Danby, Colin Rubenstein, and Robert Manne to Dennis Bark and Owen Harries, eds., *The Red Orchestra,* vol. 3, *The Case of the Southwest Pacific* (Stanford: Hoover Institution Press, 1989).

23. The Russian Orthodox church was admitted into the WCC in 1961. By the time the WCC held its conference in Vancouver in 1983, delegates from the Eastern bloc and the third world outnumbered delegates from the West by three to two. Many of the Western delegates as well held views more favorable to the Soviet Union and to third world radicalism than to the Western democracies. See Ernest W. Lefever, "Backward, Christian Soldiers," *The National Interest,* Winter 1988/9, for a recent account of the political views and activities that dominate the WCC.

24. See George K. Tanham, "Subverting the South Pacific," *The National Interest* (Spring 1988).

25. See Hayden's opening address to the Conference on Security and Arms Control in the North Pacific at the Australian National University, Canberra, in August 1987.

26. See Michael Easson, "Labor and the Left in the Pacific," in Bark and Harries, *The Red Orchestra.*

27. See Malcolm McIntosh, *Arms across the Pacific* (New York: St. Martin's Press, 1987), pp. 127–29, for a brief account of the results of American nuclear testing in the Pacific. McIntosh is a reputable English authority on Soviet and Pacific affairs.

28. *Regional Security Developments in the South Pacific.* A minority staff report to the Committee on Foreign Affairs, U.S. House of Representatives, March 1989, p. 13.

29. The New Zealand government has pledged not to promote its policies among the island states. However, the staff report to the U.S. House of Representatives notes (p.17) that some Australian and American officials "feel that New Zealand has not lived up to its commitments," and that "both in public forums and in separate political-military consultations with the island states, New Zealand officials have held up their country's anti-nuclear policies as a model for emulation."

30. See *Pacific Island Monthly,* July 1988, and *Washington Pacific Report* of 1 June, 15 June, and 1 July 1988.

NINE

Salvaging the Remnants of ANZUS: Security Trends in the Southwest Pacific

William T. Tow

D avid Lange's resignation as New Zealand's prime minister on August 7, 1989, culminated five years of historic change in alliance politics in the Southwest Pacific. Lange was elected to office as head of a New Zealand Labour government in July 1984. Since then, the tripartite Australian–New Zealand–United States (ANZUS) alliance has deteriorated into two separate bilateral defense arrangements: a close bilateral relationship between the United States and Australia and a revival of the old World War II ANZAC (Australia-New Zealand Canberra treaty) relationship between Australia and New Zealand.

Nearly all of Washington's formerly close defense ties with New Zealand are on hold until that country's laws prohibiting nuclear-armed or nuclear-capable ships or aircraft from entering New Zealand territory, put into effect nearly three years after the New Zealand Labour government assumed power, are rescinded or significantly amended. The United States does not want a repeat of the USS Buchanan episode, which occurred in February 1985. At that time, the Reagan administration challenged New Zealand's antinuclear posture by requesting that this nuclear-capable destroyer be allowed to dock in Auckland following an ANZUS maritime exercise. New Zealand responded with a request that the United States state clearly whether or not it was carrying nuclear weapons. The United States demurred, interpreting Wellington's request as a direct refutation of its "neither confirm nor deny" (NCND) policy on nuclear-armed and nuclear-capable military ships and aircraft. The incident precipitated a downward spiral in

United States–New Zealand defense relations that culminated with the American announcement in August 1986 that it no longer considered New Zealand a formal ally.[1]

The Bush administration initially hoped that Geoffrey Palmer, Lange's successor, would reconsider negotiating with United States officials on his country's ban on American military units' entering New Zealand on the basis of a "neither confirm nor deny" formula. Domestic political sentiment in New Zealand has, however, thus far continued to support the antinuclear ban. Palmer has instead called for a change in the American attitude toward the ANZUS dispute and pledged to continue the ban.[2] Even the opposition National party, which opposed the ban throughout much of Labour's two terms of office, has recently shifted toward a posture more closely aligned with that of the government's in the hope of regaining power in the next national election, which must be held no later than August 1990.[3]

Some consolation for the demise of ANZUS was found by United States policy planners in initial efforts by Australia and New Zealand to strengthen Antipodean (Australian/New Zealand) conventional deterrence capabilities in an increasingly turbulent South Pacific. More recently, however, there have been increased strains in defense relations between Canberra and Wellington. New Zealand's growing economic problems have led it to vacillate on an original Defense Ministry commitment to purchase at least two—and preferably four—West German-designed and Australian-built frigates to strengthen its South Pacific patrolling capabilities. Australian Prime Minister Bob Hawke and Defense Minister Kim Beazley warned the Palmer government that its failure to purchase at least two ships would adversely affect relations between the two countries. Opposition to the frigate purchase among New Zealand Labour parliamentarians and peace groups has, however, remained widespread.[4] They argued that New Zealand would do better to purchase smaller ships from an Asian manufacturer, which would be less expensive than relying upon Australian shipyards for construction of the frigates and which could be configured to operate at least two heavy-lift helicopters. The frigates' opponents also contended that such ships could be highly vulnerable to missile attack, especially without adequate air support, which New Zealand could not provide. Proponents of the frigates option, however, countered that no other means exists for projecting naval offensive power into key chokepoints surrounding New Zealand: the sealanes of communications transversing the Coral Sea and Solomon Islands, which could be used to interrupt vital trade and commerce if controlled by a hostile power.[5] Responding to intensified Australian pressure to complete the deal, the New Zealand Labour government affirmed its intent to move ahead with the purchase in September 1989. United States defense planners cannot take much comfort from the prospect that, as new regional economic and political nerve centers emerge in the Asia-Pacific region, Australian and New Zealand military forces are not being effectively coordinated.

American interests in the Southwest Pacific include maintaining peace in the region and working with emerging South Pacific nations and microstates to strengthen their economic prosperity and political stability. The Bush administration also wants to deflect what it regards as a disturbing trend of increasingly sophisticated Soviet diplomacy in the region. The United States strategy in this regard is designed to balance the preservation of a global deterrence posture and an increased sensitivity toward South Pacific economic and security priorities.

Importantly, from the vantage point of those advocating a continued American adherence to extended deterrence strategy, the bilateral United States–Australian defense relationship must be preserved and strengthened as a key component in United States South Pacific strategy implementation. American access to Australian naval and air bases, training facilities, and a variety of joint defense facilities located in Australia is essential to the United States' strategic command and control structure. The loss of New Zealand as an operative American ally, while regrettable, is largely manageable from the United States perspective. Australia's departure from the Western alliance system, however, would seriously damage the entire United States global collective defense system.

The Southwest Pacific's Threat Environment

The likeliest indigenous Pacific threats to Western strategic interests in the South Pacific are possible disruptions of fragile economies, local ethnic strife escalating to civil wars, and continued aid dependency, which could be exploited by opportunistic external powers. In March 1989, the Australian Parliamentary Joint Committee on Foreign Affairs, Defense and Trade released a white paper on Australia's relations with other states in the South Pacific region. The white paper underscored and documented the argument that *domestic* threats to established regimes constituted the most likely challenge to the South Pacific's overall regional stability. Conflicts materializing from external competition over political control or resource access were deemed to be far less likely to occur. New Zealand's 1987 *Review of Defense Policy* refrained from publicly identifying civil disturbances in the South Pacific island-states as a threat to that country. However, the interdepartmental Domestic and External Security Committee, which prepared the review, also generated a lengthy classified document that assessed how political instability in the South Pacific could undercut Wellington's strategic interests.[6]

A new generation of emerging Pacific island-state leaders is clearly intent on widening their range of economic and political contacts beyond traditional American, French, and Antipodean ties. However, most remain comfortable with the regional geopolitical status quo. Pacific islanders believe that, despite what they perceive to be the Western industrial countries' track record of insufficient political attention and economic aid to the region, American and Western allied policy makers are still more prone to and more capable of responding to their concerns than are other outside powers who have recently have come to view the South Pacific as strategically important. Some South Pacific political factions are becoming increasingly confident of their ability to manage relations with both superpowers and other external parties through adopting more visibly non-aligned foreign policies. Nevertheless, most island-state leaders still place greater emphasis on overcoming their region's vulnerability through economic and commercial relations with the West. They are well aware that the Melanesian, Micronesian, and Polynesian island chains are among the most aid-dependent areas in the world. They know that alternative Soviet, Chinese, Libyan, or Southeast Asian sources of support for their national and regional development aspirations cannot begin to match traditional funding by the United States, the European Community (primarily Britain, France, and West Germany), Australia, New Zealand, and, more recently, Japan.[7]

A number of South Pacific states face political turbulence that could flare up into crises with regional implications. A newly elected coalition government in Fiji was overthrown by a military coup in May 1987. Ethnic tensions between indigenous Fijians and the island's powerful but underrepresented Indian population have intensified. In Vanuatu, deep rifts between factions in the leadership group have recently spilled over into violent incidents, portending future instability. New Caledonia has posed problems as well, as native Melanesians battle loyalists who desire France to continue ruling the territory. Recent French efforts to finesse a political settlement by scheduling a referendum for New Caledonian independence in 1998 have not completely defused the issue. The other Melanesian states have formed the so-called "Melanesian Spearhead Group" to pressure for greater French concessions. Papua New Guinea has experienced an increasing number of bitter leadership changes precipitated by "no-confidence" motions in its parliament and by disputes over how to handle the land rights dispute on Bougainville that have closed the country's largest and most lucrative mine. Incessant political strife in this state, with by far the largest population in the South Pacific (3.6 million, followed by Fiji's 740,000), will continue to be an issue of great importance to Australia, which has maintained extraordinarily close economic and defense ties with Papua New Guinea.

The Soviet Dimension

Concern about South Pacific microstate development politics is shared by all three ANZUS powers. This concern represents a substantial change from the perceptions of both Australia and New Zealand during the previous decade, that Soviet activities in the region were of greatest concern. In 1976, initial—and exceedingly modest—Soviet overtures to Tonga for fishing rights and airport construction contracts were viewed by the governments in Australia and New Zealand as tactics to gain intelligence footholds and eventual military access to a South Pacific site proximate to the core areas of ANZUS operations. Canberra's and Wellington's concerns were compounded a year later with the Carter administration's failure in negotiating an Indian Ocean naval arms limitation treaty, and neither ANZUS ally felt any compunction in emphasizing South Pacific threat scenarios to press their own strategic interests with Washington's policy makers.[8]

More recently, under the Reagan administration the United States was the ANZUS member most concerned about long-term Soviet intentions in the South Pacific region. This was true even though the Soviet Union has been increasingly careful in recent years to avoid projecting a militaristic image there. Washington believes that the Soviet commercial probes and diplomatic affinity with the area's growing nuclear-free zone movement will erode the South Pacific nations' belief that their peaceful regional environment has been a byproduct of United States global military power. A recent assessment of Soviet tactics to gain incremental political advantage by aligning the Soviet Union with the South Pacific's antinuclear and intensely nationalistic political coalitions was made by Karl D. Jackson, Special Assistant to the President for National Security Affairs. It is representative of ongoing concerns of the Bush administration about the eroding credibility and relevance of United States deterrence postures directed toward the subregion.

> It is ironic, but nonetheless true, that the benign strategic environment supplied by global deterrence is little understood by some of its chief beneficiaries. Somehow the United States must communicate this point more clearly to the islands of the South Pacific. . . . The strategic equation of the Pacific remains an important component of the overall global balance, and the advocates of nuclear-free zones are deceiving themselves if they think that inky blots or parchment bonds can substitute for a stable strategic balance in the nuclear age.[9]

The Soviets have clearly stepped up their commercial and diplomatic penetration into the South Pacific region during recent years. As early as 1978 they negotiated a fisheries access agreement with New Zealand. By 1984, Moscow was offering tuna fishing rights packages to several Pacific island governments dependent on this economic sector for their national livelihood. Soviet negotiators subsequently reached successive year-long agreements for fishing trawler access with Kiribati (August 1985) and with Vanuatu (January 1987). Each deal cost the Soviets about $1.5 million and constituted substantial portions of both microstates' national income for the years they were in effect (Kiribati, for example, derived 12 percent of its total annual budget from the Soviet deal). Tonga, Tuvalu, and Western Samoa were offered similar fishing deals by Moscow but rejected them.

The Soviet Union simultaneously has moved to change its political image in the South Pacific from that of a predatory and opportunistic outsider to that of a benign external power, ready to pursue intermittent and small-scale commercial ventures. A growing number of Pacific island-states inclined to deal with the Soviet Union could justify such ties as a positive sign of their own diplomatic maturity and more independent international status. If so, low-key Soviet diplomacy in the South Pacific may well produce some immediate, if limited, political dividends, including more extensive diplomatic and consular representation in the area, consistent with Moscow's determination to gain access to all parts of the world as an international superpower.[10] To this end, Soviet diplomats, including Foreign Minister Eduard Shevardnadze and Deputy Prime Minister Vladimir Kamentsev, conducted several well-publicized tours throughout the South Pacific region during 1987–88. The Soviet ambassador to Australia also visited Papua New Guinea's capital city of Port Moresby to negotiate an agreement for a permanent Soviet resident mission in that city.

Initial fears entertained by the ANZUS powers that Soviet military and intelligence capabilities would soon follow Soviet fishing trawlers into the South Pacific, however, have thus far proven ill founded. Neither the Kiribati or Vanuatu agreements were renewed, for example, because the actual volume of the tuna catch was far too small to justify the amounts paid for fishing access rights (the Soviets lost $18 million in revenues on the Kiribati deal alone). Moscow's inability to compete economically over the long term with the United States and the other Western powers throughout the South Pacific was demonstrated when Washington negotiated a comprehensive fisheries treaty, effective in June 1988, which commits it to pay various South Pacific Forum states $10 million annually for tuna fishing rights for five years. This treaty largely negates the political showcase value of the Soviets' earlier arrangements with Kiribati and Vanuatu. Kiribati prohibited the Soviet fishing boats shore access altogether, and Vanuatu allowed only eight Soviet fishing vessels limited entry into its major towns of Port Vila and Santos. Any military-related data and intelligence the Soviets may have acquired was thus minimal. The Soviet merchant marine normally operates in Northeast Asian waters rather than in the South Pacific. Its general unfamiliarity with the vast tropical oceans surrounding these Melanesian outposts became glaringly obvious to Western analysts during the lifespan of the two agreements.[11] Disappointing volumes of tuna catches by Soviet ships during 1985–1986 belie a Soviet ignorance of tropical fisheries with their unique series of troughs, trenches, and underwater peaks on the ocean floor, a knowledge of which could otherwise be applied to anti-submarine warfare missions during wartime contingencies.

If the Soviets' true strategy in the South Pacific is the ultimate denial to the United States of control over regional sealanes and political influence in this region, it has not been very successful to date. Under the leadership of Mikhail Gorbachev, Soviet diplomacy throughout the entire Asia-Pacific has been mostly directed toward reaching accommodation with the United States, China, and (to a more limited extent) Japan. Increasingly preoccupied with his own domestic political challenges, Gorbachev has apparently not been inclined to contest the ANZUS powers' traditional footholds in the South Pacific. Soviet interests in and behavior toward the outlying Pacific islands cannot be disassociated from the reality that Moscow's greater policy concerns lie in Northeast and even Southeast Asia. Overall expansion of trade and political relations between the Soviet Union and the South Pacific island-states has remained minuscule. It appears unlikely that Soviet policy planners will assign the South Pacific region a high priority any time soon.[12]

Even the results of Moscow's obvious efforts to exploit sharpening divisions between the United States and antinuclear forces in the Southwest and South Pacific have been mixed. The ANZUS dispute, for example, has not significantly altered New Zealand's fundamental pro-Western orientation. Visiting Soviet Deputy Foreign Affairs Minister Mikhail Kapitsa discovered this when the Lange government abruptly rejected his August 1986 offer of exchanging naval intelligence information.[13]

The Soviet position on the Treaty of Rarotonga (August 1985), also known as the South Pacific Nuclear Free Zone (SPNFZ) Treaty, is also illustrative. The Reagan administration had argued that a United States ratification of the treaty's protocols was not possible because it would imply an American willingness to renegotiate long-standing NCND arrangements in NATO Europe and Northeast Asia. The Soviets moved quickly to ratify SPNFZ. Moscow, however, attached unilateral caveats to its endorsement, most notably insisting that it would not be bound by the Treaty of Rarotonga's protocols if *any* of the treaty signatories allowed United States nuclear-capable ships or aircraft access to its territory. The Soviets modified their position only when several of the SPNFZ Treaty adherents subsequently questioned Moscow's motivations on the issue, undermining the political advantages the Soviets expected to reap at the United States' expense.[14]

Other External Opportunists

Other potential strategic "opportunists" in the South Pacific include Libya, Cuba, and the People's Republic of China.[15] Libyan financial and military assistance has been linked to the radical Front Uni de Liberation Kanak (FULK) in Vanuatu, and FULK cadres were reportedly sent to Libya for advanced training in late 1984 and again in 1987. The Libyans have also developed ties with the Union Progressiste Melanesiènne in New Caledonia.

In 1988–1989, however, the "Libyan card" became less attractive to Vanuatu's prime minister, Father Walter Lini. He has been embroiled in a bitter political feud with archrival Barak Sope, a figure long suspected of cultivating Libyan ties. The Libyans attempted to project an increasingly moderate image throughout 1989, partly as a result of their de facto reconciliation with the French socialist government over Chad but more directly because the need to resolve their own internal economic problems has worked to downgrade the priority of exporting their revolutionary politics abroad.

In late 1985, Sope and other Vanuatu radicals attempted to arrange for the island-state's police force to be trained by Cuban military officers. Lini has, instead, opted for Australian and New Zealand training assistance. Earlier ANZUS concerns that Vanuatu—emerging as an independent and potentially volatile political actor on the South Pacific scene during the mid-1980s— was a prime candidate to become the "Grenada of the Pacific" have thus far proved unfounded.[16]

The People's Republic of China has become active in establishing ties with Fiji, Western Samoa, and Papua New Guinea.

Initially, Chinese motives were less opportunist and more defensive. Beijing believed it could play a legitimate defensive role in helping Australia and New Zealand to blunt the effects of the Soviet Union's establishing of diplomatic relations with Fiji in 1974. The Chinese took seriously subsequent Australian and New Zealand warnings about the expected Soviet naval expansion into distant Pacific locales. More recently, China has wanted to counter whatever political capital Taiwan might gain from its commercial ventures in Nauru, the Solomon Islands, and Tonga, all of which have formal diplomatic relations with Taipei. The last Chinese visitor of note to the area was then-Chinese Communist party Secretary Hu Yaobang in 1985. Aside from complicating the ANZUS dispute by mistakenly asserting that his country had already negotiated an NCND compromise with the United States prior to scheduled American naval port visits to China, Hu appeared to have no specific political agenda in mind. As Richard Herr has concluded, "China appears to be content largely to monitor regional events rather than to actively pursue a program of achieving foreign policy objectives [in the South Pacific].... Chinese diplomacy has created the capacity for access but to date has not sought to exploit this access to any significant degree."[17]

The South Pacific States and Western Powers: Forging Common Security Interests

If the Soviet threat or the even more distant Libyan, Cuban, and Chinese connections to South Pacific island security have subsided during the late 1980s, prospects for *Western* power intervention in the subregion are still matters of significant concern to a number of South Pacific island-state elites.

The 1987 Fiji coup was a case in point. What appeared in Canberra and Wellington to be a prudent act of placing their forces on alert in case the evacuation of their nationals from that country was required was interpreted in Vanuatu and in other Melanesian countries as the framework for unwarranted military intrusion into what they believed was strictly a domestic political crisis.[18] Pacific islanders support the concept of a tacit United States/Australian security guarantee for their sovereign integrity.

They are determined, however, to increase their influence over how and to what degree Western containment policies will be extended on their behalf.

In this context, two key strategic issues have surfaced: the role of nuclear deterrence in the subregion and the South Pacific micro-states finding an acceptable balance between defense self-reliance and strategic affiliation with the ANZUS powers.

Nuclear Deterrence and Nuclear Aversions

The South Pacific Forum membership consists of Australia, the Cook Islands, the Federated States of Micronesia, Fiji, Kiribati, the Marshall Islands, Nauru, Niue, Papua New Guinea, New Zealand, the Solomon Islands, Tonga, Tuvalu, Vanuatu, and Western Samoa. A legacy of postwar nuclear testing by the great powers in the subregion (the United States in the Marshall Islands, Britain in Australia, and France in French Polynesia) undoubtedly inspired the Forum to begin negotiating a nuclear-free zone treaty in 1983, leading to the Treaty of Rarotonga two years later. The *ideal* of a denuclearized region is universally accepted by the Forum states, but the *practicality* of continued alliance with the United States and Australia is still understood and endorsed by most of the SPNFZ Treaty signatories. While initially lobbying for a more comprehensive enforcement of antinuclear stipulations, New Zealand eventually compromised with Australia, Tonga, and Fiji, all of which insisted that the United States strategic deterrent could not be affected by the SPFNZ Treaty. Neverthless, the Lange government remained firm in its insistence that bilateral United States–New Zealand alliance relations would need to accommodate Wellington's own pending antinuclear legislation (the Nuclear Free Zone, Disarmament, and Arms Control Act passed in June 1987).

The predominant but often unspoken premise underlying the South Pacific's antinuclear movement may be characterized as a fear of "guilt by association." The premise is, if New Zealand and other South Pacific states are presumed by the nuclear-armed opponents of the United States and Australia (which hosts United States military operations directly relevant to nuclear warfighting) to be affiliated with the West during a time of nuclear war, such an opponent would have "no choice" but to deal with the South Pacific affiliates as enemies and use them as possible "nuclear pawns." The nuclear pawn logic is that select small allies of the United States could be attacked with nuclear weapons as a demonstration to Washington's other friends that they should defect from alliance with the United States in a global war environment.

It became clear by the 1980s that the Soviet Union would suffer more politico-strategic setbacks than gains by isolating and attacking small states during times of nuclear crisis. ANZUS, and its implied extension to the South Pacific, was then criticized as an unwarranted "lightning rod" that could pull the region into a nuclear war by the chance that United States or allied nuclear-capable ships and planes would be in the area. Current United States strategic doctrine, however, calls for American warships to be far out to sea and fairly distant from South Pacific littorals during the initial stages of any superpower nuclear confrontation emanating from political crisis in Europe, the Middle East, or Northeast Asia. Under such conditions, little rationale exists for any Soviet attack against New Zealand or other South Pacific nations. Soviet attacks against United States command and control operations at Australia's military facilities, of course, are much more plausible. If they occurred, New Zealand would be committed as a formal ANZUS ally to "consult" with United States and Australian officials on how to respond. The automaticity of New Zealand involvement in United States nuclear strikes against Australia's aggressor under such conditions is by no means guaranteed.[19]

The remaining, and most salient, questions are (1) to what extent New Zealand's departure from the American network of extended deterrence erodes the overall credibility of that network and (2) whether or not, as Geoffrey Palmer has asked, any such erosion matters at a time when historic breakthroughs in Soviet-American strategic arms reductions appear imminent.

The ANZUS Dispute and the United States Deterrence Network

New Zealand has shifted its strategic priorities away from the forward defense of Southeast Asia, which dominated its defense policy during the 1950s and 1960s, and toward the South Pacific. It has removed its training battalion, part of the Five Power Defense Arrangements (FPDA), from Singapore. It has reallocated its limited manpower base toward buildup of a Ready Reaction Force (RRF), committed funds to purchase frigates for enhancing its

South Pacific naval patrolling capabilities, and upgraded its military-related assistance and training programs to the Cook Islands, Niue, Tokelau, Papua New Guinea, Tonga, and other South Pacific states.

Under normal circumstances, the United States would welcome New Zealand's upgraded South Pacific strategic policy as a logical and necessary contribution to the West's overall defense burden. Pacifist elements within the Labour party and among various political interest groups in New Zealand, however, have succeeded throughout much of the 1980s in disassociating New Zealand's *national* security agenda from the regional and global deterrence strategy of the Western powers. They have insisted that New Zealand's—and, by extension, the South Pacific's—strategic environment must be regarded apart from United States deterrence strategy.[20]

In early November 1989, U.S. Secretary of State James Baker and Secretary of Defense Richard Cheney rejected an Australian declaration that the ANZUS dispute had lasted too long and should be resolved. They reiterated the long-standing American position that Washington would not "establish a separate policy" to permit New Zealand to resume its alliance with the United States while retaining its nuclear ban. They stated that the nuclear deterrence strategy was still the cornerstone of United States efforts to preserve "the peace and security of the free world." Nuclear bans and nuclear-free zones were not helpful, American officials argued, in the United States' projection of credible force throughout critical sealanes of communication or as an integral part of its nuclear deterrence posture.[21]

President Bush thus rejected arguments initially advanced by some of Michael Dukakis' campaign advisers. They had suggested that New Zealand's absence from ANZUS produced "propaganda bonanzas" for the Soviets, created frictions with Australia and the South Pacific states, and encouraged antinuclear sentiment in the region to the extent that United States capabilities to project even conventional military power in the region could be eroded.[22]

In reviewing the issue of reciprocity in the ANZUS relationship, United States force planners found that the argument often advanced by alliance theorists writing on "collective goods" (that is, Mancur Olsen and Richard Zackhauser), that small allies always receive larger payoffs than their stronger collaborators, certainly held true in the ANZUS case.[23] Throughout the life of its postwar alliance relations with New Zealand, the United States never had military communication, tracking, weapons storage, or other basing facilities on New Zealand soil. This highlighted, from the American vantage point, the selfishness underlying the Lange government's position: a fear that United States nuclear capabilities deployed in the South Pacific could pull New Zealand into an extra-regional conflict even as Wellington insisted on continuing to enjoy the benefits of a conventional deterrence affiliation with the United States. In fact, New Zealand's challenge to American alliance policies has driven Washington to reassess downward the value of New Zealand's defense contributions. The United States has proceeded to make the adjustments necessary to keep its South Pacific deterrence strategy intact in compensation for New Zealand's ANZUS departure. The first step was to systematically discontinue well over fifty bilateral or multilateral agreements linking New Zealand's military establishment with United States military personnel and providing Wellington special access to United States intelligence and weapons systems. New Zealand's access to intelligence information derived from the American-British-Canadian-Australian (ABCA) service-level information exchange programs has also been restricted. In discontinuing these arrangements, American officials concluded that they provided New Zealand "with access to information and influence in Pentagon policymaking greater than that provided for in formal treaties and also more than senior American officials felt appropriate."[24]

As the ANZUS dispute intensified, American military planners moved, in conjunction with their Australian counterparts, to modify coverage of Southwest Pacific and eastern Indian Ocean zones of commercial and naval activity previously allocated among all three ANZUS allies proportionately, under the 1951 Radford-Collins Act. The United States expanded its defense and economic assistance programs to South Pacific island-states (such as the U.S. Army's Expanded Relations Program's civic action teams' recent construction projects in the Federated States of Micronesia) in ways explicitly designed to advertise its ability and willingness to do so *independently from* similar New Zealand efforts in the region.[25] Joint United States–Australian military exercises such as the annual "Kangaroo" war games have also become more extensive in their bilateral format following New Zealand's exclusion from such maneuvers in 1985.[26] At the same time, New Zealand Ready Reaction Force exercises were becoming increasingly mired down in logistical swamps heretofore unknown in an ANZUS context and by peace activists' interference in their administration.[27] While Australian and New Zealand intelligence-sharing and maritime surveillance of South Pacific locales continued, the bulk of Australian defense and intelligence activities were diverted to meet their responsibilities in the reshaped United States–Australian bilateral alliance.

New Zealand has gradually become isolated from its other traditional Western allies, save Australia, as a result of United States efforts to isolate the ANZUS dispute from nuclear strategy in NATO Europe and elsewhere throughout the Western alliance system. This was particularly true in 1984–1985, when the United States was intent on persuading the Netherlands and Belgium to allow United States nuclear ground-launch cruise missiles on their territories as an offset to the Soviet Union's deploying SS-20 intermediate-range nuclear missiles against NATO. Washington was initially concerned that the "Kiwi disease" could create "ripples" in NATO's central and northern flanks if European peace groups seized upon the ANZUS precedent as a rationale for disrupting their countries' alliance commitments.

> The visible materialization of a "ripple" or "spillover" factor into other United States security relationships, in either Asia or Europe, which may emerge from New Zealand's present determination to raise alternatives to deterrence strategy, is a dreaded prospect for United States security planners....From a regional disarmament advocate's point of view, the interrelationship between New Zealand's refusal to allow a potential foreign (not just United States) naval unit to dock in its ports and Belgium's or the Netherlands' decision to accept to reject deployment of American nuclear deterrence forces is only an indirect one. United States' global strategy, on the other hand, requires unquestioned American access to Pacific air and naval bases at *all* levels of operations, nuclear as well as conventional, in order to reinforce adequately Europe and the Middle East during wartime.[28]

Britain (with the Royal Navy's own NCND policies) supported the United States position on the ANZUS port entry. It discontinued its own naval visits to New Zealand after the Lange government put into effect its antinuclear legislation (though continuing to interact with New Zealand military units in FPDA exercises held in the Malaysian peninsula).[29] The United States has also continued to back France in its conduct of underground nuclear tests in French Polynesia (incurring Australian as well as New Zealand's wrath on this issue). Japan, under successive Liberal Democratic party (LDP) governments, has continued to finesse the nuclear issue in its alliance with the United States by choosing not to question the content of United States naval vessels and aircraft but "assuming" that the United States "respects" its three non-nuclear principles. While the Japanese Socialist party advocates a more rigid antinuclear posture, its leadership recently has been moving closer to the LDP approach.

Regional Arms Control

Geoffrey Palmer's assertion that because the United States is negotiating with the Soviet Union it should resume defense-related contacts with New Zealand as an "old and trusted ally" fails to take into account the widening gap between United States and New Zealand positions on arms control. New Zealand, for example, has urged the United States, Britain, and France to adopt a "no first use" posture in their nuclear deterrence doctrines. All three of these powers, however, argue that nuclear reliance must remain an essential part of their own and of the West's overall strategy for maintaining escalation control against Soviet and potential third world adversaries.

New Zealand, along with Australia, has attacked the same three powers for not signing SPNFZ protocols. Bush administration officials maintain that nuclear-free zones are "not very useful" in contributing to regional stability or in ensuring that their stipulations will actually be observed by all external powers.[30] The United States instead adheres to its traditional "dual approach" of sustaining adequate regional defense while simultaneously pursuing regional detente and arms control measures. United States officials believe that pursuing arms control patiently, and from a position of strength that includes a credible posture of nuclear deterrence, will enable them eventually to achieve arms control breakthroughs in the Pacific region similar to the 1987 Intermediate Nuclear Forces agreement negotiated when the Soviets realized they could not divide the NATO European allies from the United States on this issue.[31]

Reconciling Local and Global Security Agendas

For the vast majority of South Pacific island-states, "security" cannot be viewed in the traditional sense of comparative military power and defense politics. Their concept instead revolves around how legitimate and how enduring their internal political institutions are at a given time, and around the economic progress needed to ensure continued political stability. Reconciling this viewpoint with the preoccupations of the external powers with strategic control of, or access to, trans-Pacific maritime routes and inter-regional sea and air lanes of communication will be the major challenge confronting all parties for the duration of this century and well into the next.

Thus far, the South Pacific has been relatively conflict-free compared to most other third world regions. This may change with the assumption to power of new generations of leaders who have less affinity to their countries' former colonial metropoles. Their security dilemma, however, is that they lack the necessary levels of indigenous national resources to advance their own sovereign interests and simultaneously to escape dependency on states outside the region.

The South Pacific Forum has served as the primary means of political organization and expression in the South Pacific region. The Forum Secretariat (formerly the South Pacific Bureau for Economic Cooperation, or SPEC) headquartered in Suva, Fiji, coordinates economic research and joint economic policy for its members. The Forum itself has allowed the smaller states to express their concerns about the two traditional security issues which are most important to its members—French nuclear testing and decolonization—to the great powers via Australia and New Zealand, the two "middle powers," who are Forum members as well. Because of direct Australian and New Zealand involvement in Pacific island affairs, the Forum has developed an institutional identity and policy agenda that are more pro-Western than might otherwise have been the case.[32] To what extent New Zealand's distancing itself from ANZUS will transform this relatively smooth process is still unclear.

More reassuring from the Pacific island-states' perspective is that traditional Western donors' overseas development assistance (ODA) programs to the region have continued at a fairly steady pace. Such assistance is essential. The South Pacific's island economies suffer from a lack of economic growth; island governments themselves sometimes generate over 90 percent of their countries' GNP, and private investment growth is often slow or nonexistent. Western assistance has been helpful in Vanuatu, Fiji, and other locales where domestic political turbulence has threatened to bring non-aligned regimes to power.

In this era of greater restraints on Western industrial economies, however, specific donor assistance efforts may need to be renegotiated. United States civil economic assistance to the Pacific island-states is now relatively modest, less than that of other Western donor states, including Japan, France, Australia, and New Zealand. Indeed, growing American budgetary exigencies conflict with the expectations of aid-dependent South Pacific states that Washington should pay much more than it does now for its relatively unimpeded commercial and strategic access to their subregion. The ANZUS rift, involving two of the South Pacific's most important donor nations and traditional guardians of regional stability, is a source of discomfort to the Pacific islanders. Prospects that Japan will become more involved as both an economic and a political force in the South Pacific are equally unsettling to many islanders, who have painful memories of Japanese strategic presence in the region during World War II.[33]

Economic aid and foreign investment programs provide the assistance needed for Pacific microstates to overcome diseconomies of scale. The implementation by Australia and New Zealand of mutually supportive defense policies in a post-ANZUS era offers strategic reassurance to Pacific island regimes highly vulnerable to internal change. The frigates dispute between Australia and New Zealand was an ominous development in this regard. Pacific islanders have yet to be convinced that a United States/Australia-plus-ANZAC conventional defense umbrella can really compensate for a unified ANZUS deterrent in the region. Apprehensions can only be intensified by signs that New Zealand's defense budget projections will not be sustained (a major factor behind the frigate purchase dispute with Australia), giving rise to questions about New Zealand's future willingness or capabilities to pursue any form of modified "forward defense policy." Wellington's inability to do so would not only cloud future Australian–New Zealand defense relations, but would also emit disturbing signals to the Pacific island-states. It would indicate that New Zealand, as one of the few trusted alternatives to the superpowers for economic support and political guidance, would be incapable of further sustaining its strategic influence over the wide expanse of Polynesia. This would leave Australia with even greater responsibility to maintain a strategic reach far out into the Western Pacific. What had begun for New Zealand as an idealistic campaign of nuclear dissent could deteriorate into a complete default in its responsibilities and influence as a meaningful Pacific security actor.

Conclusion

As the South Pacific geopolitical environment is increasingly shaped by local forces of change and competition, the three ANZUS signatories will be compelled to demonstrate their willingness to remain involved in the region and to contribute their resources toward ensuring regional progress and development. The United States and New Zealand, each in its own way, have been guilty of supplanting regional security interests with their intractable concerns for extended deterrence and antinuclear politics, respectively. By doing so, they have unintentionally but unmistakably undercut the cohesion and viability of Western defense strategy that has served the region so well throughout the postwar era.

As a first step toward rectifying this condition, the United States should give serious consideration to ratifying the Treaty of Rarotonga. American power projection in and through the South Pacific is unaffected by the treaty, which avoids impinging on the United States' prerogatives to move nuclear forces through the region. Washington could emphasize that *precisely because* this is the case, it can afford to respond more positively to the regional ideals and concerns expressed by the South Pacific Nuclear Free Zone than would be the case if a more literal set of treaty enforcement mechanisms were involved. Such a posture would preserve the sanctity of the NCND policy while demonstrating a renewed American sensitivity to one of the few "traditional" security positions adopted by the South Pacific Forum, a basically pro-Western organization that needs to be cultivated rather than alienated by United States policy planners whenever possible. United States misgivings about acceding to a series of nuclear-free zones that might constrain its maneuverability are not without merit. But more would be gained through this largely symbolic affirmation than lost in terms of national security.

Before any ratification took place, however, the United States would be justified in insisting that New Zealand compromise its antinuclear stance to a greater extent than it has so far been willing to do, by revising its antinuclear law to accommodate NCND on a basis of "trust but with no questions asked" (the "Japanese formula"). The United States could reciprocate by conducting what would be relatively infrequent port visits to New Zealand with naval units more clearly non-nuclear in composition than the *Buchanan*. It may well be that only a successor conservative government in New Zealand will believe it has enough political flexibility to accommodate the Americans to this extent. However, it appears that without both Washington and Wellington applying greater finesse to resolving the dispute than has been the case until present no hope exists for the eventual resolution of the ANZUS crisis. Backdoor Australian mediation, by itself, has been and will remain insufficient to achieve such a purpose.

In the absence of an ANZUS resolution, the United States, the Antipodes, and other Western powers may still be successful in ensuring that emerging South Pacific island-states ultimately will realize political stability and economic progress. Wading through the shoals of intra-regional conflict and external exploitation, however, will be more difficult than before. The price for achieving a greater measure of security in the region could be even sharper divisions over what constitutes the best approach to global security.

NOTES

1. See "Australia–U.S. Relations: Ministerial Talks — Joint Communique," *Australian Foreign Affairs Record* 57, no. 8 (August 1986), p. 741.
2. Foreign Broadcast Information Service, East Asia (FBIS-EAS) 28 August 1989, p. 60; and David Clark Scott, "New Zealand to Keep Nuclear Ban," *Christian Science Monitor*, 9 August 1989, p. 3.
3. As Thomas-Durrell Young has noted, however, it is unlikely that the United States would accept the precedent of an allied state re-entering the Western alliance with merely a change of governments, given the encouragement such a precedent would convey to antinuclear factions opposing affiliation with American extended deterrence strategy in other allied countries. See Young, "ANZUS: Requiescat in Pace?" *Conflict* 9, no. 1 (1989), p. 55. Background on the evolving National party position regarding antinuclear politics is offered by Henry S. Albinski, "American Perspectives and Policy Options on ANZUS," in John Ravenhill, ed., *No Longer an American Lake? Alliance Problems in the South Pacific* (Berkeley: Institute of International Studies, University of California, 1989), pp. 202–203.
4. FBIS-EAS-89-153, 10 August 1989, p. 75; FBIS-EAS 21 August 1989, p. 70. Also see Denis McLean and Desmond Ball, "Cooperation Enhances National Defense," *Pacific Defense Reporter* 16, no. 3 (September 1989), pp. 60–62.
5. "Questions Over Frigate Deal," *Pacific Islands Monthly* 59, no. 21 (October 1989), pp. 16–17.
6. Joint Committee on Foreign Affairs, Defense and Trade, Parliament of the Commonwealth of Australia, *Australia's Relations with the South Pacific* (Canberra: Australian Government Publishing Service, March 1989), pp. 149–50, and Steve Hoadley, "New Zealand's South Pacific Strategy," in *Strategic Cooperation and Competition in the Pacific Islands*, vol. 2 (Washington, D.C.: National Defense University, 1989), pp. 317–18.
7. Henry Albinski, "South Pacific Trends and United States Security Implications: An Introductory Overview," in R.L. Pfaltzgraff and L.R. Vasey, *Relevance of Strategic Issues in the Southwest Pacific* (Honolulu: Pacific Forum, forthcoming);

David Hegarty, "The Soviet Union in the South Pacific in the 1990s," in Ross Babbage, ed., *The Soviet in the Pacific in the 1990s* (Rushcuttters Bay, Australia: Pergamon Press Australia, 1989), pp. 123–24, and Denis McLean, "The Interests of Extra-Regional Powers," *Strategic Cooperation and Competition in the Pacific Islands,* pp. 377–401.

8. Denis Warner, ed., *Pacific Defense Reporter Yearbook 1978/79* (Church Point, New South Wales, Australia: PY Logistics and Holdings Pty, Ltd., 1979), p. 56; and Henry Albinski, *The Australian-American Security Relationship* (New York: St. Martin's Press, 1981), p. 131. Australian government views on the "Soviet threat" as it was perceived to be emerging during the mid-1970s are reflected by the testimony of Representative Philip Birney in Australia, House of Representatives, *Hansard's Daily Record,* 30th Parliament, 1st Session, 2nd Period, 20 October 1976, p. 2006. Similar New Zealand concerns are covered in "Government to Wait and See on Soviet Offers to Tonga," *New Zealand Herald,* 21 May 1976, p. 1, and "Labour Took No Action on Russians," *New Zealand Herald,* 18 June 1976, p. 1. Further background on the "Tonga episode" is offered by Hoadley, " New Zealand's South Pacific Strategy," pp. 294–97, and by Richard Herr, "Soviet and Chinese Interests in the Pacific Islands," *New Zealand Herald,* 18 June 1976, p. 325.

9. "The Red Orchestra and the U.S. Response in the South Pacific: An Exploration of Metaphor," in Dennis L. Bark and Owen Harries, eds., *The Red Orchestra: The Case of the Southwest Pacific* (Stanford, CA: Hoover Institution, 1989), p. 13. For a corroborating view, see George K. Tanham, *The Soviet Union in the South Pacific,* RAND P-7431 (Santa Monica, CA: The RAND Corporation, April 1988).

10. This point is well developed in Richard Herr, "The Soviet Union in the South Pacific," in Ramesh Thakur and Carlyle A. Thayer, eds., *The Soviet Union as an Asian-Pacific Power* (Boulder, CO: Westview Press, 1987), pp. 135–52.

11. An unverified report on Vanuatu's alleged acquiescence in the placement of Soviet acoustical tracking equipment in its waters is by Marc Liebman, "More Soviet Activity in the South Pacific," *Armed Forces Journal International* 124, no. 12 (July 1987), p. 28. Also see Herr, "Soviet and Chinese Interests," p. 333; Hegarty, "The Soviet Union in the South Pacific," pp.

117–19; and Tanham, *The Soviet Union in the South Pacific,* pp. 2–3.

12. For an opposing viewpoint, see Michael Barnard, "Soviet Try to Hook into Pacific Region," *The Age* (Melbourne), 12 July 1988, p. 13; Dr. Robert F. Miller, "Hidden Text in Soviet Policy," *Pacific Defense Reporter* 15, nos. 6/7 (December 1988/ January 1989), p. 15, who cites two Soviet analysts of Pacific affairs writing for a relatively obscure Soviet journal, *MEMO* (August 1987) that SPNFZ is the best means for denying access to the region to United States and allied forces and for ultimately luring the Treaty of Rarotonga signatories into "a far reaching general security agreement involving the USSR, giving it major influence on the parameters and conditions of stability in the region." Also see E. Rampell, "Soviet Objectives 'To Be Disruptive,'" *Pacific Islands Monthly* 57, no. 8 (August 1986), p. 22, and Bark and Harries, *The Red Orchestra: The Case of the Southwest Pacific,* passim.

13. Richard Long, " PM Rejects Soviet Information Offer," *The Dominion* (Wellington), 28 August 1986, p. 2. Most South Pacific countries, moreover, actually sided with the United States in the ANZUS dispute as the Americans took care to consult closely with a number of South Pacific leaders about the continued need for adherence to NCND as an integral part of the ANZUS security blanket, which most of them coveted. For assessments of this process, see Michael Danby, "Tales of Trouble from the South Pacific," *Asian Wall Street Journal,* 25 May 1989, p. 6, and Albinski, "South Pacific Trends and United States Security Implications," pp. 31–32 of draft manuscript.

14. The initial Soviet caveat noted that "in the event of any actions undertaken by the state or states, which are parties to the Rarotonga Treaty, in violation of their main commitments under the treaty connected with the non-nuclear status of the zone and perpetration by one or several states — parties to the treaty of an act of aggression with the support of a state having nuclear weapons or jointly with it with the use of by such a state of the territory, air space, territorial sea or archipelago waters of those countries for calls by naval ships and flying vehicles with nuclear weapons on board or transit of nuclear weapons, the Soviet Union will have the right to consider itself free from

the commitments undertaken under Protocol Two of the treaty. In the event of any other actions by the parties to the treaty incompatible with their non-nuclear status, the USSR reserves for itself the right to reconsider the commitments undertaken under the said protocol." *Pravda,* 16 December 1986, p. 5, as translated and reprinted in Foreign Broadcast Information Service, Soviet Union (Daily Report), 17 December 1986, p. E-2. Also see Jackson, "The Red Orchestra and the U.S. Response in the South Pacific," pp. 10–11; and Hegarty, "The Soviet Union in the South Pacific," p. 117. Accounts of the official United States position are by Norman Kempster, "U.S. Rejects Nuclear-Free S. Pacific Zone," *Los Angeles Times,* 6 February 1987, p. 5, and Neil A. Lewis, "U.S. Bars Plan for Nuclear Free Zone," *New York Times,* 5 February 1987, p. 6.

15. The term "opportunists" is used by Denis McLean in his "The Interests of the Extra-Regional Powers," in *Strategic Cooperation and Competition,* pp. 389–96. The discussion in this paragraph is largely extracted from McLean's analysis.

16. Robert C. Toth, "U.S., Australia Wary of Soviets' S. Pacific Role," *Los Angeles Times,* 22 February 1987, pp. 1, 14–15; Hamesh McDonald, "Checking the Soviets," *Far Eastern Economic Review* 134, no. 40 (2 October 1986), pp. 26–27; and Frederic A. Moritz, "Global Powers Vie in the South Pacific," *Christian Science Monitor,* 29 January 1987, pp. 9–10.

17. Herr, "Soviet and Chinese Interests in the Pacific Islands," in *Strategic Cooperation and Competition,* p. 328.

18. Henry S. Albinski, "Security Issues within the Pacific Island Community," in *Strategic Cooperation and Competition,* pp. 184–85, offers particularly salient analysis on this point.

19. Both the "nuclear pawn" and "lightning rod" theories are discussed in Ramesh Thakur, *In Defense of New Zealand: Foreign Policy Choices in the Nuclear Age* (Boulder and London: Westview Press, 1986), pp. 69–72.

20. The definitive statements justifying this New Zealand posture are by Prime Minister David Lange, "The Oxford Union Debate," reprinted in *New Zealand Foreign Affairs Review* 35, no. 1 (January–March 1985), pp. 7–12, and by Lange, Ministry of Foreign Affairs Press Statement, no. 7, May 1986, p. 9. Also see Kennedy Graham, *National Security Concepts of States: New Zealand* (New York: Taylor & Francis, 1989), pp. 31, 45–47.

21. Norman Kempster, "U.S. Rejects Call to Repair N. Zealand Ties," *Los Angeles Times,* 4 November 1989, p. A-3.

22. John A. Cutler, "Will U.S. Political and Military Policy toward South and Mid-Pacific Change in the Bush Administration?" *Contemporary Southeast Asia* 11, no. 1 (June 1989), pp. 81–82.

23. Mancur Olson and Richard Zackhauser, "An Economic Theory of Alliances," *Review of Economics and Statistics* 48 (August 1966), pp. 266–79.

24. Peter Jennings, *The Armed Forces of New Zealand and the ANZUS Split: Cost and Consequences,* Occasional Paper No. 4 (Wellington: New Zealand Institute of International Affairs, 1988), p. 4. Jenning's work is the most complete account of the ANZUS divestiture and its ramifications for New Zealand available to date. Also see Thomas-Durrell Young, "ANZUS: Requiescat in Pace?" pp. 52–54, and Hoadley, "New Zealand's South Pacific Strategy," pp. 302–304.

25. William Maddox, "The U.S. Army in the Pacific," *Asia-Pacific Defense Forum* 14, no. 1 (Summer 1989), p. 38.

26. See an Agence France Presse report by Ian Pedley, 1 August 1989, as translated and reprinted in Foreign Broadcast Information Service, East Asia, 1 August 1989, p. 72, for background on Kangaroo '89—the biggest joint exercise between Australia and the United States since World War II and involving some 25,000 personnel.

27. "Joint Venture '86," for example, conducted on the Cook Islands during July 1986, demonstrated the New Zealand forces' inability to counterattack against a well-armed invader due to logistical problems in arming Royal New Zealand Air Force Skyhawk fighter-bombers and refueling naval frigates. Roger Mackey, "Joint Venture '86," *Evening Post* (Wellington), 14 July 1986, p. 4. "Operation Golden Fleece," was conducted in a remote area of New Zealand's North Island and simulated a Ready Reaction Force intervention against rebel forces in a "mythical South Pacific state." It was disrupted by two self-styled "mediation groups" demonstrating against the RRF concept. See David Robie, " Golden Fleece No Force for Peace," *Pacific Islands Monthly* 59, no. 15 (March 1989), p. 25.

28. William T. Tow, "The ANZUS Alliance and United States Security Interests," in Jacob Berkovitch, ed., *ANZUS in Crisis:*

Alliance Management in International Affairs (Basingstoke and London: Macmillan Press, 1988), p. 66.

29. Dora Alves, "The Changing New Zealand Defense Posture," *Asian Survey* 29, no. 4 (April 1989), p. 372.

30. For a representative account of New Zealand's position on nuclear-free zone politics, see Kennedy, *National Security Concepts of States: New Zealand,* pp. 81–82. Until his recent retirement, Australian Foreign Minister Bill Hayden was the most forceful advocate of his nation's support for nuclear-free zone politics throughout much of this decade. An account of his opposition to United States, British, and French positions on SPNFZ is by Chris Sherwell, "Australia Says Pacific Lacks Arms Controls," *Financial Times,* 1 May 1987, p. 3. The Bush administration's position is recounted by Kempster, "U.S. Rejects Call."

31. The United States position is spelled out by then-Assistant Secretary for East Asian and Pacific Affairs Gaston J. Sigur, "The Strategic Importance of the Emerging Pacific," *Current Policy* No. 871 (29 September 1986).

32. This point is well addressed by Robert C. Kirste, "The Pacific Islands: A Contemporary Overview," in *Strategic Cooperation and Competition,* pp. 66–67.

33. Analysis of Japan's new "strategic aid" emphasis in ODA programs targeted toward the South Pacific, "Tokyo Bearing Gifts," *Far Eastern Economic Review* 135, no. 5 (29 January 1987), pp. 30–31, and in Akio Watanabe, "The Pacific Islands and Japan: Perspectives and Policies," in *Strategic Cooperation and Competition,* especially pp. 265–68.

Notes on Contributors

JUNE TEUFEL DREYER is professor of politics and director of East Asian programs at the University of Miami.

HARRY G. GELBER is professor of political science at the University of Tasmania, Australia.

LAWRENCE E. GRINTER is professor of Asian studies at the Maxwell Air Force Base, Alabama.

OWEN HARRIES is editor of *The National Interest* and former Australian ambassador to UNESCO (1982–83).

DONALD C. HELLMAN is professor of international studies and political science at the Henry M. Jackson School of International Studies, University of Washington.

RICHARD G. LUGAR (Republican-Indiana) is a member of the Subcommittee on East Asian and Pacific Affairs, United States Senate, and former chairman of the Senate Foreign Relations Committee.

EDWARD A. OLSEN currently serves as professor of Asian studies and associate chairman for research in the Department of National Security Affairs at the Naval Postgraduate School.

DOUGLAS PIKE is director of the Indochina Studies Project at the University of California at Berkeley and editor of *Indochina Chronology*.

WILLIAM T. TOW is assistant professor of international relations at the University of Southern California.

DONALD S. ZAGORIA is professor of government at Hunter College and a fellow of the Research Institute on International Change and the Harriman Institute for Advanced Study of the USSR at Columbia University.

Index

DATE DUE
